AF304882

Cold War Air Thieves

Stealing Military Aviation Technology and the East-West Battle for Aviation Secrets

Cold War Air Thieves

Stealing Military Aviation Technology
and the East-West Battle for
Aviation Secrets

Steven Taylor

First published in Great Britain in 2026 by
Pen & Sword Airworld
An imprint of
Pen & Sword Books Ltd
Yorkshire - Philadelphia

ISBN 978 1 03614 349 7

Typeset in INDIA by IMPEC eSolutions
Printed and bound in England by CPI Group (UK) Ltd, Croydon, CRO 4YY

The Publisher's authorised representative in the EU for product safety is Authorised Rep Compliance Ltd., Ground Floor, 71 Lower Baggot Street, Dublin D02 P593, Ireland.
www.arccompliance.com

For a complete list of Pen & Sword titles please contact

PEN & SWORD BOOKS LIMITED
47 Church Street, Barnsley, South Yorkshire, S70 2AS, England
E-mail: enquiries@pen-and-sword.co.uk
Website: www.pen-and-sword.co.uk

or

PEN AND SWORD BOOKS
1950 Lawrence Rd, Havertown, PA 19083, USA
E-mail: uspen-and-sword@casematepublishers.com
Website: www.penandswordbooks.com

'It is a basic principle of warfare that to know
the weapons the enemy has is already to beat him.'

*General Dan Tolkowsky, commander of the
Israeli Air Force, 1953–1958*

Contents

Acknowledgements

Firstly, I would like to thank the team at Pen & Sword Books. For their kind permission to reproduce images and quote extracts from various books and magazines, I'd like to express my gratitude to the following: Dr Paul Lashmar (Spy Flights of the Cold War), Crecy Publishing (Red Star Volume 7: Tupolev Tu-4 – Soviet Superfortress), Biteback Publishing (Mossad: The Great Operations of Israel's Secret Service), The History Press (Looking Down the Corridors), Orion Publishing Group (Wings on my Sleeve), Yale Representation Ltd (Secret Agencies: US Intelligence in a Hostile World), Oxford Publishing (The Dragon in the Jungle), Simon & Schuster (A Spy For All Seasons), the US Naval Institute (Proceedings magazine), Josef Campion (Aviation News magazine), Timothy Dowling (Journal of Military History), Petar Milošević (image of cockpit canopy from shot-down F-117) and Samuel Lynch of the National Museum of the US Air Force (cover image of Lt No Kum-sok's MiG-15).

Finally, I wish to thank my family, without whose support and encouragement nothing would get done.

Steven Taylor, October 2025
www.steventayloronline.co.uk

Historical Note

Soviet intelligence underwent multiple changes of nomenclature throughout the 69-year history of the USSR. Whatever name it went by, however, the organization's basic roles remained unchanged.

The Soviet Union's first intelligence service was the All-Russian Extraordinary Commission for Combatting Counter-revolution and Sabotage, popularly known as the Cheka. It was briefly renamed the State Political Directorate (GPU) in 1923, before becoming the OGPU the following year. The next change came in 1934 when it was rebranded the NKVD. From 1943 there was a series of further changes in nomenclature in quick succession – NKGB (1943 – 46), MGB (1946 – 53) and MVD (1953 – 54) – before settling on the name by which Soviet intelligence is best known, and which remained in place until the dissolution of the USSR in 1991: the KGB.

From the 1940s, Soviet intelligence officers engaged in the collection of Scientific and Technology (S & T) intelligence were known as Line X officers, and were stationed at so-called 'legal' residencies in the target countries, operating under official diplomatic or trade cover. There were also 'illegal' residencies, undeclared to the host nations, in which the officers had false identities. The HQ of the Soviet intelligence service was known as Moscow Centre (often referred to simply as 'the Centre'), to which intelligence reports were sent.

The Soviet military possessed its own separate foreign intelligence service, the Main Intelligence Directorate, known as the GRU. Established in 1918, this name has remained constant and continues to the present day.

Both the KGB and the GRU were engaged in gathering information on the West's military aviation during the Cold War, and there was frequent rivalry between the two organizations. This became sufficiently problematic for Moscow Centre to complain about the lack of cooperation between the two in 1963.

- Due to the multiple name changes in the 1940s and early 1950s, for reasons of simplicity I use the term NKVD when referring to the Soviet intelligence service of the 1934 – 1954 period.

Glossary

AAA	Anti-Aircraft Artillery
AAM	Air-to-Air Missile
ADD	Soviet long-range aviation (1942-1944)
AFB	Air Force Base
AFC	Air Force Cross
AI	Airborne Intercept
AMTORG	American-Soviet Trading Organization
ARCOS	All-Russian Co-Operative Society
ASA	Army Security Agency
ASW	Anti-Submarine Warfare
ATIC	Air Technical Intelligence Centre
AVMF	Soviet naval aviation
AWACS	Airborne Warning and Control System
BfV	West German domestic intelligence service
BRIXMIS	British Commanders-in-Chief Mission to the Soviet Forces in Germany
CIA	Central Intelligence Agency
Comintern	Communist International – organization created by the USSR to promote communism abroad
CPGB	Communist Party of Great Britain
DA	Long-range aviation division of Soviet/Russian Air Force (1946 – present)
DARCOM	US Army Materiel Development and Readiness Command

DARPA	Defense Advanced Research Projects Agency
DIA	Defense Intelligence Agency
DST	Directorate of Territorial Surveillance (French domestic intelligence service, 1944 – 2008)
DSTI	Directorate of Scientific and Technical Intelligence
ECM	Electronic Counter-Measures
Elint	Electronic Intelligence
EOD	Explosive Ordnance Disposal
FBI	Federal Bureau of Investigation
FCD	First Chief Directorate (foreign intelligence arm of the KGB)
'Five Eyes'	Alliance of intelligence agencies of the US, UK, Canada, Australia and New Zealand
FRG	Federal Republic of Germany (West Germany)
FTD	Foreign Technology Division
GAZ	Soviet State Aircraft Factory
GDR	German Democratic Republic (East Germany)
GKNT	Soviet State Committee for Science & Technology
GPU	State Political Directorate (Soviet intelligence service, 1922 – 23)
GRU	Main Intelligence Directorate (Soviet military intelligence service, 1918 – present)
HUR	Main Directorate of Intelligence (Ukrainian intelligence service, 1992 – present)
HVA	Foreign intelligence arm of the Stasi (1955 – 1990)
IAF	Israeli Air Force
IAP	Soviet Fighter Aviation Regiment
IFF	Identification Friend or Foe
KGB	Committee for State Security (Soviet intelligence service, 1954 – 1991)
LD/SD	Look-Down/Shoot-Down radar
LII	Soviet Flight Research Institute

LSK	East German Air Force
MfS	Ministry for State Security (East German intelligence service, better known as the Stasi, 1950 – 1990)
MLM	Military Liaison Mission
MSS	Ministry of State Security (Chinese intelligence service, 1983 – present)
NACA	National Advisory Committee for Aeronautics
NATO	North Atlantic Treaty Organization
NII	Scientific Research Institute of the Soviet Air Force
NKVD	People's Commissariat for Internal Affairs (Soviet intelligence service, 1934 – 1943)
NSA	National Security Agency
NVA	North Vietnamese Army
OGPU	Joint State Political Directorate (Soviet intelligence service, 1923 – 34)
OKB	Experimental Design Bureau
OP	Observation Post
ORB	Operations Record Book
PLA	People's Liberation Army
PLAAF	People's Liberation Army Air Force
PR	Photo-Reconnaissance
PRC	People's Republic of China
PVO	Air Defence Forces of the Soviet Union
R & D	Research & Development
RAE	Royal Aircraft Establishment
RAF	Royal Air Force
RCAF	Royal Canadian Air Force
RCMP	Royal Canadian Mounted Police
RDF	Radio Direction Finding – original British term for radar
ROCAF	Republic of China Air Force (Taiwanese Air Force)
RuAF	Russian Air Force

S & T	Scientific & Technological intelligence
SACEUR	Supreme Allied Commander Europe
SB	Polish intelligence service (1956 – 1990)
SBU	Ukrainian domestic intelligence service (1991 – present)
SHAPE	Supreme Headquarters Allied Powers Europe
SIGINT	Signals Intelligence
SIS	Secret Intelligence Service (also known as MI6)
SOVMAT	Soviet Material
StB	Czech intelligence service (1945 – 1990)
STOVL	Short Take-Off, Vertical Landing
SVR	Russian foreign intelligence service (1991 – present)
SWT	S & T collection department of the HVA
TsAGI	Central Aerohydrodynamic Institute
TsIAM	Central Institute of Aero Engines
UAV	Unmanned Aerial Vehicle
USAAC	United States Army Air Corps
USAAF	United States Army Air Force
USAF	United States Air Force
VPAF	Vietnamese People's Air Force
VPK	Soviet Military-Industrial Commission
VVS	Soviet Air Force
Warsaw Pact	Military alliance of USSR and its Eastern and Central European satellite states

Introduction

On Tuesday, 30 December 1947, Viktor Yuganov, a 28-year-old test pilot with the Mikoyan OKB, took a new Soviet fighter into the air for the first time. The stubby, silver jet with swept-back wings that took off that day from the USSR's Flight Research Institute airfield at Ramenskoye, a few miles southeast of Moscow, was the MiG I-310, the prototype of what would later be redesignated the MiG-15. With a top speed of 648 mph, a maximum ceiling of 50,800ft, and a heavy armament consisting of two 23mm and a single 37mm cannon, the MiG-15 represented the cutting-edge in jet fighter design. Three years later, the MiG-15 would become an aviation legend in the skies over North Korea, feared and respected by the airmen of the United Nations who had to face it in combat.

But this marvel of the Soviet aviation industry owed more to German and British technical know-how than Soviet. The swept-wing configuration was heavily influenced by Nazi research captured at the end of the Second World War, while the powerplant that allowed the MiG to achieve its impressive performance, the Klimov RD-45, was a pirated copy of the British Rolls-Royce Nene gas-turbine engine.

The dramatic intervention of the MiG-15 in the Korean War, which completely changed the complexion of the air war during that conflict, prompted an enormous effort by the British and Americans to acquire a flyable example of Stalin's new superjet. After successfully obtaining wreckage of MiG-15s in several daring missions behind enemy lines, an intact example eventually fell into US hands, thanks to a North Korean defector.

The epic saga of the attempts by Western intelligence to obtain a MiG-15 for detailed analysis and extensive flight testing – what the Americans termed FME (Foreign Military Exploitation) – was just one of many secret operations carried out during the Cold War, by both sides, to steal aviation technology. Some of these operations are relatively well-known; many others have remained almost unknown to this day.

Gaining access to, and unlocking the secrets of, the enemy's latest warplanes and other aviation-related hardware, whether it be radar, ECM equipment or weapons systems, was one of the most important areas of intelligence collection during the long, tense stand-off between East and West known as the Cold War. Had the Cold War turned into an all-out shooting war between NATO and the Warsaw Pact, the side possessing the greatest knowledge of its opponents' military hardware – and especially its aircraft – would, of course, enjoy a crucial advantage that may well have proven decisive. According to one CIA analyst: 'Foreign-material collection played a great role in our ability to make accurate assessments of performance – the real capabilities of systems – and often monitor the changes that were made.'[1]

For the USSR, there was a secondary – and perhaps even more important – objective in stealing the details of the latest aviation projects of its adversaries. Lagging behind the West in most areas of technical development, industrial espionage offered the Soviets a quick and cost-effective means of bridging the technological gap, shaving years and millions of roubles off the development process.

The Soviet Union's campaign of aviation espionage began as soon as the new communist state came into being in 1922, following the Russian Civil War. Even when the Soviet Union found itself in alliance with the chief targets of its espionage activities, Britain and the United States, during the Second World War, Moscow's intelligence-gathering activities continued. During the Cold War, these activities expanded. 'A formidable apparatus was set up for scientific espionage; the scale of this structure testified to its importance,' stated Gus Weiss, a White House policy adviser specializing in technology and intelligence.[2]

Thanks to its vast intelligence apparatus and the appeal communism held for many idealistic people in the West, allowing the NKVD and its successor organizations to recruit willing spies at the heart of the European and North American aircraft industries, Moscow was able to greatly enhance its technical capabilities in many key areas. From strategic atomic bombers to the gas-turbine engine and heatseeking air-to-air missiles, espionage was critical in allowing the Soviet Union to keep pace with the West. 'Soviet intelligence was professional at ferreting out science and technology and had the results to prove it. The Soviets were adept at copying foreign designs,' wrote Weiss.[3]

Indeed, the Soviets became heavily dependent on 'reverse-engineering' Western aircraft. Though there was no shortage of talent in the Soviet aircraft industry, purely homegrown designs often fell victim to the prejudices of Soviet leaders, those of Josef Stalin in particular. Behind the propaganda that extolled the inherent superiority of the communist Soviet system over the 'decadent and corrupt' Western capitalist structure, he privately considered Western – especially American – military hardware and engineers superior to those of the USSR.[4] A classic example of this attitude was the post-war project to develop a long-range strategic bomber to carry the Soviet atomic bomb, with Stalin favouring a reverse-engineered copy of the American B-29 Superfortress over the distinguished Soviet designer Andrei Tupolev's own offering, the 'Aircraft 64' project. This thinking continued throughout the Cold War. 'Soviet planners,' revealed a 1980s US intelligence report, 'use Western technology as a yardstick to evaluate their own capabilities ... and, with some important exceptions, pursue technologies already proven in the West.'[5]

By the early 1980s, the CIA reported that the Soviet intelligence effort to steal the West's technology was 'massive, well planned, and well managed – a national-level program approved at the highest party and governmental levels.'[6] The scale of this campaign of industrial espionage was detailed in another CIA report from 1985: 'Each year Moscow receives thousands of pieces of Western equipment and many tens of thousands of unclassified, classified, and proprietary documents as part of this campaign. Virtually every Soviet military research project – well over 4,000 each year in the late 1970s and over 5,000 in the early 1980s – benefits from these technical documents and hardware. The assimilation of Western technology is so broad that the United States and other Western nations are thus subsidizing the Soviet military build-up.'[7]

The Western powers, on the other hand, scornful of the quality of Soviet aircraft and the technical capabilities of its aviation industry, at first exhibited scant interest in obtaining intelligence on Soviet military aircraft. Instead, the main effort of the West's intelligence agencies was directed towards counter-espionage – preventing the Soviets from gathering intel on their aircraft, a task in which they enjoyed mixed success. Not until 1950, when the MiG-15 ended the air supremacy the UN forces had established over North Korea, did this situation change, as the West realized that they could no longer afford to dismiss the products being churned out in Soviet factories.

The closed, highly secretive nature of Soviet society, however, made the task of collecting intelligence on Soviet aircraft extremely difficult, and throughout the Cold War many of their aircraft remained objects of considerable mystery to the Free World, their capabilities the subject of heated debate within the West's intelligence community. Nevertheless, the CIA, the French DST and British intel-gathering outfits such as BRIXMIS did pull off several notable intelligence coups. Amongst these were the capture by British military personnel of components of a crashed Yak-28P 'Firebar' interceptor and the recruitment by the DST of a key figure in Directorate T, the KGB section dedicated to the acquisition of Western technology.

The secret battle for aviation intelligence fought in the shadows by the spies of the capitalist and communist countries raged throughout most of the twentieth century. But the dissolution of the USSR in 1991, marking the official end of the Cold War, did not bring this battle to a close. After a brief thaw in East-West relations in the 1990s, the arrival in the Kremlin of Vladimir Putin in 2000, with his confrontational stance towards NATO, has sent the West's relations with Moscow back into the deep freeze, and the intelligence war has become as intense as at any time during the Cold War. Meanwhile, the rise of China to superpower status in the twenty-first century has presented NATO with another – and arguably much more dangerous – potential antagonist, as the Chinese, like the Soviets before them, have sought to rapidly close the technological gap between East and West through industrial espionage, particularly in the area of military aviation.

While many books have been written about intelligence activities during the Cold War, none have been devoted exclusively to examining the largely neglected area of aviation espionage. My aim with this book is to rectify that situation, bringing to light some of the most daring intelligence missions that took place on both sides of the Iron Curtain to steal the closely-guarded secrets of some of the most potent warplanes ever devised.

Chapter One

The First Cold War

At the outbreak of the First World War, Imperial Russia possessed numerically the largest air force in the world, with 263 aircraft. However, the Tsar's Imperial Russian Air Service was something of a paper tiger, hobbled by poor serviceability and inadequate leadership. Russia's aviation industry, meanwhile, was incapable of matching the production standards of its European counterparts. Consequently, Russia was obliged to rely heavily on imported engines and aircraft from its British and French allies. From the French they received Gnome engines, Farman bombers and Nieuport fighters; from the British, the Royal Aircraft Factory BE2e and Vickers FB.19, among other types. The Russian Lebedev company also manufactured versions of the British Sopwith Tabloid and 1½ Strutter fighter under licence at its factory in St Petersburg (then known as Petrograd), powered by imported Gnome engines, though the Russian copies of these aircraft suffered from poor workmanship and used inferior materials.

Anglo-Russian cooperation also extended to the development of new aviation technology. During a visit to England in 1916 as part of a Russian mission to observe British aircraft manufacturing methods, decorated Russian naval aviator Lieutenant Victor Dybovsky joined forces with Warrant Officer F.W. Scarff of the Admiralty Air Department to develop interruptor gear (a device which allows a machine gun mounted above an engine block to fire through the propeller arc without hitting the blades).

After the overthrow of Tsar Nicholas II in 1917 by the communist Bolsheviks under Vladimir Lenin, in March 1918 the government of the new Russian Soviet Federative Socialist Republic signed a peace deal with Germany, the Treaty of Brest-Litovsk, ending direct Russian participation in the war. The armistice between Russia and Germany prompted the despatch of an Allied expeditionary force to the northern Russian ports of Archangel and Murmansk to prevent the vast stocks of armaments and supplies sent by Britain and France to the Tsar's forces from falling into German hands. After the defeat of Germany, this intervention escalated into active involvement by Allied – and especially British – forces in the ongoing Russian Civil War, who fought on the side of the anti-Bolshevik 'White' Russians.

With the bulk of what remained of the Imperial Russian Air Service taken over by the Bolsheviks, it fell to the RAF to provide the 'Whites' with much of their airpower. Among the squadrons sent to Russia were Nos 47, 221 and 266, equipped with Royal Aircraft Factory RE8s, Sopwith Camels and Pups, Bristol F2bs, Airco DH9s and Avro 504Ks.

War-weary after four years of brutal conflict with the Central Powers, British and Empire airmen in Russia, like their army and naval counterparts, suffered from low morale, not helped by the poor conditions they found in Russia. Compounding this were the chronic divisions in the 'White' camp. Unsurprisingly therefore, and despite some notable acts of bravery by individual pilots, they achieved little and by mid-1920 the Allied contingents had been withdrawn from Russia. The Civil War staggered on until 1922, with Lenin's Bolsheviks emerging victorious and formally establishing the Union of Soviet Socialist Republics.

Building an aircraft industry

The ravages of the Civil War had left the country's aviation industry in a dire state. In 1922, Soviet industry turned out a pitiful 43 airframes and just eight aero engines. With airpower gaining acceptance among the world's major military powers as potentially the dominant future form of warfare, the Soviet Government was determined to reconstruct the country's shattered industry and build up a new Soviet Air Force, known at the time as the Workers' and Peasants' Red Air Fleet.

To begin with, this meant manufacturing unlicensed copies of British aircraft that had fallen into Bolshevik hands during the Civil War. Among these were the Polikarpov R-1, based on the Airco DH.9 and built at the GAZ-1 factory in Moscow, and the U-1, a copy of the Avro 504K, an example of which was captured by the Bolsheviks when it force-landed in northern Russia. Known by the Soviets as the *Avrushka* ('Little Avro') and powered by a Russian-made copy of the French Gnome Monosoupape engine, almost 700 U-1s were built (most at the GAZ-23 factory in Leningrad, as St Petersburg had now been renamed), serving as the Soviet Air Force's standard basic trainer until 1932, along with a further 73 of a seaplane variant, the MU-1.

The Soviets soon moved on to designing and producing their own aircraft, though their early attempts met with only limited success. In 1923 Polikarpov unveiled the Soviet Union's first monoplane fighter, the I-1. But the I-1 possessed some lethal handling characteristics and only two prototypes and 33 production machines were manufactured, none of which entered service with the Soviet Air Force.

Moscow also turned to its erstwhile foe, Germany, for help in modernizing its aircraft industry and introducing Soviet designers and engineers to the latest production methods. Under the terms of the Treaty of Versailles, the new German Weimar Republic was prohibited from possessing a military air arm, with tight restrictions also placed on the production of aircraft in Germany. To circumvent these restrictions, in 1922 the Soviets invited the Junkers Aircraft and Motor Works to set up production at the GAZ-7 factory at Fili in Moscow. In return, Junkers would share their aeronautical expertise and train Soviet engineers in the construction of all-metal aircraft, an area in which Junkers was a world leader. Among the future legends of Soviet aviation who worked at the Fili facility and benefitted from this arrangement were Andrei Tupolev and Pavel Sukhoi.

Russo–German cooperation in aviation matters deepened in 1926 when a deal was negotiated to establish a clandestine fighter pilot school near Lipetsk, 270 miles southeast of Moscow. Given the cover name 'Scientific Experimental and Personnel Training Station' by the *Reichswehr* (the armed forces of the German Weimar Republic), both German and Soviet airmen were trained at the school.

With the Nazis' accession to power in 1933, relations between the two countries soured and industrial and military cooperation came to an abrupt end. That same year the Lipetsk fighter pilot school was shut down. An estimated 120 German pilots, along with more than 300 groundcrew personnel, had been trained at the school during its seven years of operation. These graduates would go on to form the nucleus of Hitler's Luftwaffe.

Josef Stalin, who had become Soviet leader following Lenin's death in 1924, was determined to build up a strong Soviet military aviation industry, one that could compete with, and hopefully one day surpass, those of the great Western powers – which he regarded as the natural enemies of the USSR and communism – as part of his grand ambition to industrialize Russia. In a speech to the First All-Union Conference of Workers of Socialist Industry in 1931, he declared: 'We are fifty or a hundred years behind the advanced countries. We must make good this distance in ten years. Either we do it, or we shall be crushed.'[1]

The Herculean effort to industrialize the USSR was to be realized through the infamous 'five-year plans', the first of which was announced by Stalin in October 1928. A sum of 426 million roubles was invested in the State aviation industry to bring it up to Western standards. By 1932 the number of major aircraft plants in the USSR had increased from 15 to 22 (with nine more under construction), while the workforce directly employed in aircraft construction skyrocketed from less than 10,000 to 84,000 over the same period.[2] By the end of the first five-year plan period, the Soviet Union was producing over a thousand aircraft per annum.

In the 1930s the Soviets produced the world's first production monoplane fighter with a retractable undercarriage, the Polikarpov I-16. Soviet airmen, flying Soviet machines, also set numerous aviation records. On 21 November 1935 Vladimir Kokkinaki set an – unofficial – world altitude record of 48,067ft in a modified Polikarpov I-15 fighter, and in July 1936 Aleksandr Belyakov, Valery Chkalov and Georgy Baidukov made the world's longest non-stop flight – at 5,825 miles – between Moscow and Udd Island (now Chkalov Island) in a Tupolev ANT-25.

These achievements enhanced the reputation of Soviet aviation and airmen, but could not disguise from foreign observers the fundamental weaknesses of the country's aviation industry, and the fact that it was failing to keep pace with technological developments in the West. The poor quality of Soviet engines, in particular, was noted by the British air attaché in Moscow in a report he sent to London in 1937. The Japanese were also left unimpressed after inspecting examples of Polikarpov fighters shot down in their border clashes with the Soviets in the late 1930s. In May 1937, an American diplomat at the US embassy in Moscow reported: 'Despite the fact that Soviet pilots, in Soviet-made planes have established a number of world records, it would appear that the military authorities at least are not satisfied with the development of Soviet aviation.'[3]

Development of the aircraft industry in the 1930s was not helped by Stalin's paranoia, which led to some of the country's leading aeronautical designers and engineers, including Andrei Tupolev, Nikolai Polikarpov and Vladimir Petlyakov, being swept up in the purges known as the Great Terror and thrown into labour camps, where they were compelled to continue their work, before eventually being 'rehabilitated' and released. Many other talented aviation workers perished in these camps.

Well aware of the shortcomings of Soviet industry, Stalin sought to close the technological gap between the Soviet Union and its Western rivals by directing his intelligence agencies to steal aviation technology and designs from the West. In 1925 his feared spy chief Felix Dzerzhinsky, head of the USSR's first intelligence agency, the Cheka, had already made S & T intelligence-gathering a key objective, with the British aircraft industry initially being the prime target of Soviet espionage.

The Arcos Affair

During the 1920s, Soviet espionage activities in the UK centred around a front organization, ARCOS (the All-Russian Co-Operative Society), run by the GRU out of the Russian Trade Delegation's offices at 49 Moorgate in Central London. One member of the ARCOS staff investigated by MI5 was Leonid Vladimiroff, an engineer whose job was to negotiate the purchase of British aircraft and aero engines on behalf of the USSR. This role allowed him to regularly visit the factories of major aviation firms, including Handley-Page and Vickers. MI5 stated in a report that he 'boasts of his absolute power to enter the most secret workshops and drawing offices of these firms.' But MI5 failed to establish Vladimiroff's direct involvement in espionage. Nevertheless, they concluded: 'Subject's contacts in the aircraft development and manufacturing world must have been of interest to the GRU'.[4]

MI5 enjoyed more luck with their investigation into Wilfred Macartney. Born in Fife, Scotland, Macartney served with distinction in the First World War, being awarded a Certificate of Merit for making a daring escape from his German guards after being taken prisoner in 1918, and also worked for a time as an intelligence officer in the Middle

East. Post-war, he became involved in Left-wing politics, joining the CPGB and writing for the socialist *Sunday Worker* newspaper.

Recruited by the OGPU, Macartney was tasked by his handler Georg Hansen – known to him as 'Mr Johnson' – with acquiring aviation intelligence, who provided him with a detailed questionnaire on the RAF and British military aviation. To get the answers, he approached George Monckland, a friend and colleague at Lloyds Bank, where Macartney worked and whom he believed had contacts at the Air Ministry. Spooked by Macartney's request, Monckland secretly alerted the authorities and a full-scale Security Service investigation was set in train, headed by Guy Liddell. Through Monckland, MI5 fed Macartney material on the RAF that would be of little real value, including an aircrew training manual that was due to be replaced, while mounting round-the-clock surveillance of Macartney and the ARCOS premises.

Having established the involvement of ARCOS staff in espionage, a major raid was mounted on their offices by MI5, City of London police and armed Special Branch officers on 12 May 1927. Two Russians were caught destroying incriminating files, though enough material was recovered to prove conclusively that the Soviets were involved in illegal spying activities in the UK. Later that month, the British Government used this evidence to justify breaking off diplomatic relations with the USSR, which had only been established three years before.

Macartney, however, was kept in play. Believing he had avoided suspicion, he continued trying to secure intelligence from Monckland to sell to the Soviets. 'On the closing down of Arcos, Macartney found himself without a job, and, being desperately in need of money, and finding that he could obtain what were he thought correct answers to the questionnaire through Monckland, has been endeavouring to get into touch with the Soviet organizations abroad with a view to selling his wares. At the same time he has been posing to Monckland as persona grata with the Soviet organization with a view to obtaining more information. This he wished to hand over to the Soviet S.S. [Secret Service],' MI5 reported.[5]

As the investigation progressed it became clear to Liddell and his team that an entire spy ring of British communists was active, engaged in gathering economic and military intelligence for Moscow, which MI5 dubbed 'the Kirchenstein Organization', after the British traitors' handler, Jacob Kirchenstein, the OGPU's head of station in Britain. A key figure in this spy ring was James Messer. A transport engineer and known communist agitator from Glasgow, Messer was described in his MI5 file as 'a notorious and dangerous revolutionary' who was 'deeply implicated in Soviet espionage plans.' In 1926, Messer arrived in London to organize a branch of the CPGB for spying. MI5 later reported that Messer and his group of sub-agents were gathering 'information re the new designs going through the Ordnance Department of Armstrong Whitworth's, and the new designs of aircraft manufactured by A.V. Roe.'[6]

Finally, in November 1927, MI5 made their move. Macartney and Hansen were arrested, the former being charged with offences under the Official Secrets Act, including 'attempting to obtain information on the RAF.' Their trial was held at the Old Bailey in January 1928. Unconvinced by Macartney's defence that he was merely a freelance journalist trying to obtain scoops, the jury found both men guilty and each was sentenced to 10 years in prison. Hansen was later deported to the USSR.

No sooner had the ARCOS spy ring been broken up than more British communists suspected of passing aviation intelligence on to Moscow came to the attention of the Security Service. One of these was Thomas Mercer, a journalist who worked as assistant editor on a technical journal and one-time member of the CPGB. He first came to the attention of the security services in 1928, a Special Branch report dated 21 January stating that 'according to an informant,

MERCER was gathering information on Royal Air Force matters and was also interested in the mechanical sections (tanks etc.) of the Army. MERCER was said to be canvassing Communists who had been connected with the Air Force for information.'[7]

Communist infiltration of the RAF was a source of growing concern to MI5 at the time, the Service reporting in February 1928 that there were believed to be 'several [Communist] Party members in the RAF and there is a National Minority Movement Group at [RAF] Kidbrooke Aerodrome.[8] Those concerned are now attempting to secure "contacts" in the aircraft factories, ostensibly to find out about working conditions, but really to try and obtain information regarding machines.' By 1926 Mercer had apparently left the CPGB, but MI5 believed he remained a communist sympathizer and possible Soviet agent. 'He is in the category of people who are more valuable to Moscow outside than inside the Party,' they reported.[9]

Also attracting MI5's interest was Mercer's girlfriend Susan Benda, a fellow journalist and active communist who, according to Special Branch, was 'one of the leaders and contacts in this country of the "Eyes and Ears of the Soviet", the name given to a section of the Russian espionage service.' She was also suspected of gathering intelligence pertaining to British aircraft. 'During the 1914–18 war she had been on the staff of an air paper: she was believed to have a thorough knowledge of aeroplane construction and motors,' Special Branch reported.[10]

Intermittent surveillance of the pair continued over the next few years and in June 1932 Mercer was observed by Special Branch detectives at the Hendon Air Display, in company with another suspected communist, 'taking what appeared to be detailed photographs of the new types of aircraft displayed in a special enclosure.' However, Special Branch was unable to establish if either Mercer or Benda were directly engaged in espionage on behalf of the OGPU, admitting in a 1933 report that 'it has not been possible by enquiry to connect

[Mercer] recently with any overt extremist activities.'[11] And so the Security Service wound down its investigation into the pair.

As the 1930s progressed, further cases of suspected communist espionage within the British aircraft industry came to light. In 1936 an anonymous tip-off led MI5 to Eric Camp, a draughtsman at the Gloster Aircraft Company. When he was subsequently arrested by Special Branch, they found a note on him on which was scrawled performance statistics of a monoplane fighter Gloster were developing to meet Air Ministry Specification F.5/34, which was on the Secret List, along with details of the Bristol Mercury X engine. Camp insisted the statistics were simply to assist him with a project he was working on in his spare time for a revolutionary remote-control aircraft gun-aiming system. But as the note was addressed to Mikhail Sokoloff, manager of the Engineering Division of ARCOS (which had resumed business in London in 1932) and a suspected OGPU intelligence officer, MI5 was understandably sceptical of his explanation.

Charged with breaching the Official Secrets Act and 'making notes calculated to be, or intended to be, useful to an enemy, for a purpose prejudicial to the safety or interests of the State', Camp was remanded to HMP Brixton. His trial was held *in camera* at the Old Bailey in October 1936. Camp protested his innocence. 'I was working on secret aircraft and I asked, quite openly, for information from a colleague that would help me with my invention. He gave it and I wrote it on a slip of paper and put it in my wallet. That is my only crime,' he told the court.

Giving evidence for the prosecution, Captain Norman Liptrot of the Air Ministry asserted that 'these performance figures would be of great value to a foreign power.' His defence counsel, however, explained away Sokoloff's name and the address of ARCOS on the note by telling the jury that Camp felt his lack of a degree had held back his career in the UK and so was seeking employment with the Soviets. 'He thought that in Russia, where less importance was attached to this matter than in England, he would have a better opportunity of improving his position.'

The prosecution was unable to convince the jury that he was passing military secrets to the Soviets and he was instead found guilty of the lesser charge of being unlawfully in possession of classified material. Camp was spared jail and instead bound over for two years good behaviour. In a press interview after the trial, Camp insisted he was 'absolutely loyal to my King and country, and have never at any time dreamt of betraying secrets to an enemy.' However, during one of the monthly meetings with a Special Branch officer he was obliged to attend as a condition of his probation, the officer recorded that Camp spoke of his communist leanings. This officer added that he judged Camp to be 'wholly untrustworthy where the safety of the State is concerned.'[12]

The Gloster fighter project, meanwhile, was cancelled after two prototypes had been built, the Air Ministry instead selecting the Hawker Hurricane and Supermarine Spitfire as the RAF's standard day fighters.

Communist spies at the RAE

The most serious case of Soviet aviation espionage in Britain during this period involved a suspected spy ring operating within the Royal Aircraft Establishment. Based at Farnborough airfield in Hampshire, the RAE was Britain's main aeronautical research and testing facility for new aircraft and aviation technologies, and therefore a prime target for Soviet intelligence.

In the mid-1930s, MI5 was receiving disturbing reports from a source at Farnborough that 'an active and dangerous Communist cell' was operating at the RAE, the leader of which was believed to be Wilfred Foulston Vernon. A pilot in the Royal Naval Air Service in the First World War and later the RAF, after the war Vernon became the chief draughtsman for the Bristol Aeroplane Company before joining the RAE in 1925, eventually becoming one of its senior technical officers. Described in one report as being 'fanatically devoted to Communism', by 1935 his outspoken Left-wing political views and a trip he made with several RAE colleagues to the USSR caused MI5 to regard him as a potential security risk. The Security Service raised its concerns about Vernon with the RAE's management, recommending that he be dismissed. But their recommendation was rejected as it was felt that 'the case against Vernon was hardly strong enough for them to dismiss Vernon without giving him any reasons.'[13]

MI5's investigation fizzled out owing to a lack of evidence. But the case was revived when his house was burgled in August 1937 and, after the thieves were arrested, the police found classified RAE documents they'd stolen from his house, including blueprints of the Avro Anson transport aircraft and notes on new fighters and bombers entering service with the RAF, including the Fairey Battle, Gloster Gladiator and Armstrong-Whitworth Whitley. Though unable to prove Vernon was supplying this material to a foreign power, he was charged with improper possession of secret documents, fined £50 after pleading guilty, and dismissed from the RAE.

Among other technicians at Farnborough suspected of being members of a pro-communist clique were Ben Lockspeiser and Frederick Meredith, both of whom accompanied Vernon on his trip to the USSR. Lockspeiser had joined the RAE in 1921 and would go on to become one of the country's most eminent scientists. His flirtation with communism in the 1930s attracted the attention of the Security Service. However, no evidence that he was involved in passing on secrets was unearthed and, despite being periodically investigated by MI5 over the following years, his career was unaffected. Knighted in 1946, he worked on Britain's atomic weapons programme in the 1950s and eventually became head of the Department of Scientific and Industrial Research.

Meredith, a senior technical officer at Farnborough, was thought to be more deeply involved. Born in Killiney, County Dublin, Meredith was considered one of the RAE's most brilliant brains, who helped develop early auto-pilot systems and radio control target drones, such as the de Havilland DH.82 Queen Bee. But his communist views alarmed the Security Service, with their informant at Farnborough branding him 'one of the most dangerous men employed at the RAE.'

In 1934, MI5 held a meeting with senior Air Ministry officials to discuss the potential security risk posed by Meredith. They found the Ministry unwilling to take any action against him, as the scientist's skills were 'of such a high order that the Air Ministry were prepared to run very considerable risks in order to retain his services.' His employers further argued that 'if Meredith were dismissed, he might quite possibly go to Russia where the Air Ministry would be completely deprived of his services and brain, which they value very highly.'[14] As with Lockspeiser and Vernon, MI5 was forced to drop its investigation into Meredith for lack of evidence.

And there the matter might have rested, had it not been for the chance discovery of the so-called Robinson Papers. This was a collection of over 800 secret papers collected by Henri Robinson, a Belgian communist and GRU section leader covering Switzerland, France and the UK, which was discovered by British forces in a house during the Allied advance through France in 1944. The papers led MI5 to Ernest Weiss, a German Jewish émigré and one of the GRU's top spy handlers in Britain prior to the Second World War. When he was finally interrogated by MI5 in January 1948, Weiss named Vernon and Meredith as two of his most important assets, and confirmed that they had been recruited in 1936. He also confirmed that Vernon had given him the blueprints of the Avro Anson, while Meredith supplied details of the Queen Bee target drone, among other things. This 'was perhaps the most interesting information disclosed by him,' MI5 noted. Nevertheless, Weiss revealed to his interrogators that his superiors were not satisfied with the intel produced by his agents. 'Weiss has stated that he was frequently taken to task by his spymasters for the poor quality of the information and material supplied by Major Vernon and Meredith,' they reported.

Armed with this new evidence, Meredith was interviewed by MI5 officer Jim Skardon, by which time he was working at Smiths Instruments on missile guidance systems. Under interrogation, Meredith admitted his espionage activities, insisting he had acted out of ideological motives and had broken off contact with the Soviets in 1939. Skardon thought Meredith a somewhat reluctant spy. 'The only document that Meredith purloined from RAE was [a] lantern slide of the schematic of an automatic bomb sight. This disclosure by Meredith caused him the greatest personal anxiety,' his report of the interrogation revealed. Skardon's report continued: 'From time to time, Meredith would be asked questions [by Weiss] of a purely military character, e.g. the dispositions and formation of squadrons in the RAF, but he always refused to provide the answers.'[15]

Despite his confession, the authorities chose not to prosecute Meredith for espionage, judging his skills as the country's leading authority on missile guidance systems indispensable to the British defence industry. Nevertheless, concerns over Meredith's loyalty persisted, with MI5 seeking assurances from his employers that efforts would be made to limit his access to secret information.

The investigation into Wilfred Vernon was complicated by political factors, as he was now serving as a Labour Member of Parliament. The British Prime Minister Clement Attlee was briefed on Vernon's pre-war involvement with the Soviets in May 1948. It was now clear that his possession of secret documents in 1937, for which he had been fined, 'was part of a far more sinister design than was at the time capable of proof,' MI5 told the PM.[16] But it was agreed not to interview Vernon. Only after losing his parliamentary seat in the 1951 general election was he finally interrogated by MI5, during which he confessed to working for the Soviets. No prosecution resulted as it was felt too much time had elapsed since the offences had taken place.

As relations between London and Moscow sharply deteriorated after the signing of the Molotov-Ribbentrop Pact in August 1939 and Stalin's invasion of Finland in November that year, MI5 uncovered yet another Soviet attempt to elicit intelligence from British aviation workers, including plans for the new Rolls-Royce Merlin engine.

In November 1939 Alexei Doschenko, head of the Engineering Department of the Soviet Trade Delegation in London, made contact with an engine fitter known to him as 'Matthews', who worked at Rollasons Aircraft Services – which overhauled RAF aircraft – and told the Russian he was willing to sell military secrets to the USSR. Doschenko instructed him to obtain information on the armament of British fighters, bomb-release mechanisms and the Merlin engine.

After a meeting with Matthews on 20 December 1939, Doschenko was arrested by Special Branch detectives at Chancery Lane tube station. 'Matthews' was in fact Thomas Hendrick, a Rollasons employee who had been encouraged by his works manager to infiltrate local political activists suspected of trying to spread communist propaganda amongst the company's workforce. Special Branch also found evidence on Doschenko indicating that he had been trying to obtain information on the latest catapult system the Royal Navy was developing to launch aircraft from the decks of ships.

Protected from prosecution by his diplomatic status, Doschenko was deported from the UK in January 1940 for 'endeavouring to collect secret information by illegal means', his explusion prompting an official protest from the Soviet ambassador Ivan Maisky.

The deportation of Doschenko was another setback for Soviet aviation intelligence operations in the UK. However, the Soviets would soon gain some first-hand experience of the Merlin engine, thanks to their short-lived alliance with the Nazis. In March 1941, the Luftwaffe invited Soviet test pilot Stepan Suprun to try out a Spitfire Mk Ia in Germany, which they'd captured the previous year. Suprun carried out two flights and his impressions of Britain's most illustrious wartime fighter were generally favourable, praising its stability and ease of operation, though he was critical of the Spitfire's short endurance and lack of heavy calibre machine-guns.[17]

Stalin's American comrades

By the 1930s, Moscow's aviation espionage effort was increasingly concentrated on the United States, the OGPU appreciating the technical superiority of the country's industry. 'Nowhere is technology as advanced in every sphere of industry as in A. [America] … This makes tech. intelligence in the USA the main focus of work,' Moscow Centre instructed its spies in 1934.[18] Another advantage of operating in the US was that, with no dedicated domestic counter-intelligence outfit, American industry was relatively easy for Stalin's spies to penetrate. The nearest equivalent the United States had to MI5 was the FBI, which at that time was primarily geared towards investigating organized crime.

One of the first Soviet operatives sent to the United States to carry out industrial espionage was Abram Einhorn, who arrived in New York in 1930 posing as a businessman. Codenamed 'Taras', Einhorn remained in the States until early 1934, during which time his most notable achievement appears to have been acquiring details of a diesel aero engine being developed by the Packard company. Einhorn was later swept up in the mass arrests of the Great Terror and sentenced to eight years imprisonment.

Moscow Centre also devised a brazen new strategy for gathering technical intelligence in the US: enrolling intelligence officers as students at America's most prestigious universities, to study subjects including engineering, chemistry and aeronautics. Not only would these student officers benefit from a first-class education, greatly improving their own knowledge which could then be passed on to Soviet educational establishments, but they would also be in an ideal position to identify and recruit promising American students in the campuses sympathetic to the communist cause as agents.

In the summer of 1931, Moscow Centre dispatched a cadre of 75 Soviet students to New York. Among the universities they would attend were Harvard, Cornell and Stanford, with the largest group – 25 in all – enrolling at the Massachusetts Institute of Technology. This group was led by one of the Soviets' most important intelligence officers of the interwar period, Stanislav Shumovsky, who enrolled in a three-year undergraduate course in aeronautical science at MIT.

Born into a middle-class family in Zhytomyr, Ukraine in 1902, Shumovsky trained as a pilot in the mid-1920s and served in a reconnaissance squadron of the Red Air Force. But his flying career was cut short by a crash in his Polikarpov R-1 that left him with an injured left arm. After studying aeronautics at Moscow's Bauman Institute, where one of his teachers was the legendary Soviet aircraft designer Andrei Tupolev, he was recruited as an intelligence officer by the OGPU and given the codename 'Blériot', after the pioneering French aviator Louis Blériot.

At MIT, Shumovsky recruited several important assets among his fellow students. One of these was Benjamin Smilg, the son of Jewish Russian émigrés, whom Shumovsky had befriended when Smilg gave him private tutoring to improve his English. According to his NKVD file, Smilg – assigned the codename 'Lever' – had been recruited by Shumovsky in 1934. After graduating from MIT, he went to work for the Glenn L. Martin Aircraft Company, before securing a post in the research department at Wright Field, where new aircraft were tested and aviation technology developed. This, stated the NKVD, made Smilg a prize asset, as 'Wright Field employees are supplied with plenty of information regarding developments in aviation in the USA and other countries.'

Industrial rather than State espionage carried out by business competitors was the main security concern at Wright Field, Smilg told Shumovsky, who reported back to Moscow Centre:

> ...because WF [Wright Field] is a repository of secrets from various aviation companies and there is a fear that WF employees might give one company's secrets to another, all employees – especially civilians – are kept under constant surveillance. They are categorically forbidden to meet with representatives from other companies, especially outside the walls of the establishment. Surveillance is set up so well that, no sooner has someone dined with a company representative, than the next day he is summoned by his boss for questioning and disciplining.[19]

NKVD files suggest that Smilg supplied Shumovsky with information on aero engines and aerodynamics, including how to solve the problem of vibrations – or 'flutter' – affecting the structural integrity of bomber tail assemblies. Smilg proved a troublesome and half-hearted agent, however. In 1937 he stopped passing on information to Shumovsky. When he eventually came under suspicion and was questioned by the FBI in 1950, Smilg denied having ever spied for the Soviets and, lacking hard evidence, the FBI could bring no charges of espionage against him.

Intelligence-gathering in the States became even easier with the establishment of formal diplomatic relations between the USSR and the US in 1933. 'Legal' Soviet residencies were set up in New York, Washington DC, San Francisco and Los Angeles, which became centres of Soviet espionage, their staff enjoying some protections under the rules of diplomatic immunity. In 1933 Gaik Ovakimian (codename 'Gennady'), a qualified chemist and OGPU officer known to the FBI as 'the wily Armenian', took up a post as an AMTORG official at the New York residency. His actual role, as Deputy Head of the Scientific-Technical Section, was to supervise the S & T collection activities of the Soviet spies infiltrated into America, including Shumovsky.

After graduating from MIT in 1934, Shumovsky also went to work as a representative of AMTORG, which afforded him regular access to US aircraft plants, ostensibly to inspect new designs for possible purchase for the Soviet Air Force but in reality to identify and recruit more sources among the workforces of America's leading aviation firms.

In 1935, his mentor Andrei Tupolev arrived in the US, heading a delegation of Soviet engineers to aircraft factories to select products for possible purchase, which would then be built under licence in Soviet factories. Although denied access to the more sensitive aeronautical technology under development at these facilities, Tupolev wrote afterwards that he gained 'a lot of valuable information' during his visit. The designer was particularly struck by the fact that, at a time when Soviet and European manufacturers were still producing large numbers of biplane fighters, their American counterparts were focusing exclusively on monoplanes.[20]

The factories of the Douglas Aircraft Corporation proved to be especially fertile recruiting ground for Shumovsky. At the company's El Segundo plant in California, he recruited a draughtsman, an engineer and a specialist in high-altitude flight, codenamed, respectively, 'Falcon', 'Gapon' and 'Tikhou'. None of these agents has ever been identified.

His prize catch, though, was Jones Orin York, who was given the codename 'Needle'. A qualified pilot and engineer, York was recruited by Shumovsky in 1935 while working at Douglas. Questioned by the FBI many years later, he admitted that the Soviet engineer had expressed interest in the design of a new aero engine he was working on and gave him $200 'with the idea that eventually I might develop something that the Russians would actually purchase.' A few months later, he told his interrogators, 'Shumovsky asked me to furnish him information from Douglas Aircraft Corporation, El Segundo Division, which I did.' York also told the FBI that he was motivated by financial reward rather than ideological conviction.[21]

The NKVD was extremely pleased with the quality of the intelligence they were receiving from York. '"NEEDLE" has a wonderful attitude towards his work,' a report enthused. 'He carries out all of our assignments with precision and care. He is extremely happy with the work and has repeatedly expressed his warm feelings with regard to the Soviet Union.'[22] In 1937, Moscow Centre proposed setting up a special technical committee dedicated exclusively to evaluating Needle's material. However, the upheavals within Soviet intelligence brought about by the Great Terror ultimately scuppered this plan.

By 1939 Soviet S & T intelligence-gathering in the US had reached enormous proportions. In that year alone, 18,000 pages of technical documents, many pertaining to aviation, were forwarded to Moscow Centre. In recognition of this success, Ovakimian was promoted to Chief of Scientific Intelligence in the US.

But just as it was reaching its apogee, Soviet aviation espionage suffered a series of setbacks. In February 1939, Shumovsky was recalled to Moscow to take up a new role as Deputy Chair of the Technical Council, People's Commissariat for Aviation Industry. His inexperienced replacement in America, Semyon Semyonov, was, initially at least, far less effective at recruiting and handling new agents.

Soviet operatives in America also found themselves under much greater FBI scrutiny, following the uncovering of a Nazi spy ring active at the USAAC's Mitchel Field air base on Long Island, New York a few months earlier. This led to tightened security at military installations and aircraft factories across the country, and prompted the FBI to make counter-espionage a top priority. Suspecting he was under FBI surveillance, in 1940 York broke off contact with his Soviet handler. Increased FBI vigilance paid off in May 1941 when Ovakimian, stressed and overworked, was caught in a sting operation while receiving classified documents from an agent in a New York restaurant, dealing a major blow to Soviet intelligence operations in the States.

Compounding these difficulties was the signing of the Molotov-Ribbentrop Pact in August 1939, effectively turning the USSR and Nazi Germany into allies, which disillusioned some of their more ideologically motivated American agents, many of whom ended cooperation with their handlers.

But in June 1941, dramatic events on the other side of the world would completely transform the situation, throwing an unexpected lifeline to Soviet intelligence and leading to a golden period in espionage, in both Britain and the US.

Chapter 2

The Uneasy Alliance

In the early hours of 22 June 1941, 138 German divisions smashed through the Soviet Union's western frontier. Operation Barbarossa, Hitler's invasion of the USSR, had begun. Putting aside his lifelong ideological opposition to communism, Churchill immediately pledged full and unequivocal British support to the Soviet Union, offering to 'give whatever help we can to Russia and the Russian people' in a national radio broadcast hours after the start of the German invasion.[1]

This assistance initially came in the form of a British military mission under Major General Noel Mason-Macfarlane, known as 30 Mission, which was despatched to Moscow just two days later to liaise with the Soviets and coordinate supplies of British military materiel to Russia. The Mission also had a Naval and Air Section, the latter headed by Air Vice Marshal Sir Conrad Collier, who had served as the air attaché at the British Embassy in Moscow in the 1930s.

With the Soviets losing over 1,200 aircraft on the opening day of Barbarossa alone (most on the ground), replacement fighters from Britain were Moscow's most pressing requirement. Churchill responded by shipping large numbers of Hawker Hurricanes, mostly Mk IIBs, to the port of Murmansk, the first batch arriving in September. A detachment of RAF pilots and groundcrew, known as 151 Wing, accompanied the first Hurricane deliveries to help integrate the new type into service with Soviet Naval Aviation. To augment the northern sea route, which was vulnerable to German U-boat and air attack, a second, overland supply route to southern

Russia from the Middle East was established. This required a joint Anglo-Soviet invasion and occupation of Persia (now Iran) in August, with the RAF and Soviet Air Force quickly sweeping aside the small Imperial Iranian Air Force.

Opinions of the Soviet pilots who flew the Hurricane in combat on the Eastern Front were decidedly mixed. While Senior Lieutenant Vitaly Klimenko dismissed the Hawker machine as 'a piece of junk', the highly decorated ace Lieutenant Colonel Boris Safonov, who headed the Soviet training programme on the British fighter, declared that he 'loved the Hurricane'.[2]

From late 1942, Spitfires from Britain also began to arrive in the USSR. The Soviet attitude towards the legendary Supermarine fighter was generally more positive. Though lacking the robustness of the Hurricane, Soviet pilots appreciated its excellent all-round performance. In addition to deliveries of aircraft, a group of Soviet airmen were brought to Scotland in early 1943 to be trained on the Armstrong-Whitworth Albemarle torpedo-bomber, a dozen of which were then supplied to the Soviets but used only in the transport role.

The United States added to this bounty when it extended its Lend-Lease scheme – which provided war materials to the UK on a loan and return basis – to the USSR in October 1941, an arrangement which would see an eventual total of over 14,000 aircraft delivered to the Soviets, including Bell P-39 Airacobras and P-63 Kingcobras, North American B-25 Mitchells and Douglas A-20 Havocs. The USAAF, however, took the precaution of first removing some of the more sensitive equipment, such as the Norden bombsight, from the donated aircraft.

British and American generosity in providing vast materiel aid to the USSR earned little gratitude from Stalin. Nor did it buy much co-operation in the exchange of intelligence and technology. Traditional Soviet secrecy and Stalin's deep-rooted mistrust of Western motives ensured the flow of intelligence was strictly one way, with 30 Mission and its US equivalent, established in October 1943 and headed by Major General John Deane, granted little insight into Soviet military hardware or the course of the air war on the Eastern Front. General Mason-Macfarlane and AVM Collier's frustration with this state of affairs was revealed in a telegram sent to the War Cabinet on 3 November 1941, in which they complained that 'co-operation with the Russians was becoming extremely difficult and that the Mission were unable to collaborate technically.'[3]

This would be a familiar Allied complaint throughout the war. At a Cabinet meeting held on 18 February 1943, the Secretary of State for War Sir James Grigg advised his colleagues that the Chiefs of Staff 'could not tolerate the continuance of the present position, in which we communicated a great deal of information [to the Soviets] and got nothing in return.'[4]

Due to this secrecy, the Western Allies' knowledge of Soviet aircraft remained extremely sketchy throughout the war. The prevailing view among senior British and American officers was that Soviet designs and production standards fell well below those of the West. This attitude was reflected by Major General Deane, who in his post-war memoirs wrote dismissively that '[Soviet] industry is dependent on foreign machine tools, and unless assisted by foreign technical advice it is stumbling and inefficient. My experience and observations convince me that the Russian people simply do not have industrial "know how" comparable to ours'.[5] Aside from the volunteer French pilots of the Normandie-Niemen squadron, who flew the Yak-9 fighter on the Eastern Front, only in the immediate aftermath of the war did a very few airmen of the Western Allied nations get a brief opportunity to fly Soviet types, thanks to informal contacts between junior officers.

One of these was the famed British test pilot Captain Eric 'Winkle' Brown. As the CO of the RAE's Enemy Aircraft Flight, whose job was to test fly and evaluate captured enemy aircraft, he was sent to Germany in the summer of 1945. One of the facilities he visited was

Tarnewitz, a Luftwaffe weapons testing site on the Baltic coast in Soviet-occupied eastern Germany. Relations with the Soviets were by this time already beginning to unravel, but Brown was fortunate in that the Russian commander of the site was reasonably well disposed towards the British, having flown Hurricanes provided by the UK early in the war. This officer granted Brown permission to try out two of the Soviets' most successful wartime aircraft, the Yakovlev Yak-9 fighter and the Petlyakov Pe-2 bomber. Brown was reasonably impressed, regarding both as 'pretty well up with the state of the art', though he was critical of the standard of the cockpit instrumentation.

On a later visit to the facility, he managed to secure test flights in two more workhorses of the Soviet fighter fleet, the MiG-3 and Lavochkin La-7. 'The MiG-3 was old hat stuff and not very impressive,' Brown wrote in his memoirs. But he thought more highly of the newer La-7, which entered service in 1944. 'Its handling and performance were quite superb, and it had the qualities necessary for a fine combat fighter,' he wrote. On the other hand, he was critical of the La-7's wooden construction and considered it under-armed.[6]

The observations of Soviet aircraft by Brown and others led the RAF's Air Intelligence Branch to conclude in a post-war report that the Soviet aircraft industry suffered from 'a shortage of skilled workers.' The report continued: 'The fighter aircraft are well finished externally and the machining and finish has been of a high order where efficient functioning of the equipment has demanded it. But where roughness of finish cannot affect efficiency, it is as though the village blacksmith had been called in to help.'[7]

Radar and 'Window'

Radar (known by the British during the war as Radio Direction Finding, or RDF) was one technology which the Soviets were keen to acquire. Some important work had been done on radar in the USSR in the pre-war years. But the combination of the Great Terror of 1937–39, which saw leading Soviet radar engineer Pavel Oshchepkov sent to the Gulag for ten years, and the dislocation to industry caused by the Nazi invasion had severely retarded Soviet development in this field. Therefore, in early 1942 the Russians made a request through 30 Mission for ASV (Air to Surface Vessel) radar, for use in aircraft to seek out and destroy German U-boats. This presented the British with a dilemma: such equipment was top secret, and there were concerns that the Russians couldn't be relied upon to protect the secrets of British radar. On the other hand, refusal risked damaging the Anglo-Soviet alliance. The chairman of the RDF Policy Sub-Committee, Sir Henry Tizard, 'considered that ASV would be of very great assistance to [the Soviets] in the Baltic and Black Seas, in particular in the destruction of enemy submarines', and his recommendation was that the sets be handed over to the Russians.[8]

But senior RAF officers, including the Vice-Chief of the Air Staff, Air Chief Marshal Sir Wilfred Freeman, argued the risks were too great. 'Though he saw no reason to doubt the effectiveness of the Russian security measures, he was not in favour of giving details of our ASV equipment to Russia because there was a considerable risk that these sets might fall into enemy hands through forced landings,' minutes of a War Cabinet meeting held on 16 March recorded.[9] The Americans were also opposed to the transfer of ASV radar technology to the Russians, on similar grounds.

In the end, Churchill decided that maintaining good relations with Moscow trumped the security concerns of the RAF top brass and backed the recommendation of the RDF Policy Sub-Committee. In April 1942 a team of RAF radar specialists was sent out to Moscow, taking with them radar equipment to instruct the Soviet Air Force in its use. The British also gifted the Russians examples of its AI Mk IV radar, for use in the air-to-air interception role. The AI Mk IV is believed to have influenced the development of the Soviets' first indigenous AI set, the Gneiss-2. Installed in a small number of twin-engine Petlyakov Pe-2s, these constituted the VVS's first radar-equipped night-fighters, making their combat debut in late 1942 before being officially approved for service in summer 1943.

Across the Atlantic, Soviet spies were also hard at work stealing the secrets of American radar. The Soviet effort to acquire Allied radar technology through espionage was given the codename Project *Raduga* ('Rainbow'). Two of their most important agents in this field were electrical engineers Alfred Sarant and Joel Barr – codenamed, respectively, 'Hughes' and 'Meter' – who worked for Western Electric. Sarant and Barr sent over 9,000 pages of documents relating to US radar systems to the Soviets.

When Britain gained the USSR as an ally in the war against Nazi Germany, Churchill had ordered the British intelligence services to desist from all espionage activities directed against the Soviet Union. This gesture was not, of course, reciprocated by Stalin, who ordered his network of spies in Britain and the US to get to work stealing the secrets of the aviation technologies Churchill and Roosevelt refused to freely disclose to the USSR.

The man put in charge of gathering S & T intelligence in the UK was Vladimir Barkovsky. Born in Belgorod in 1913, Barkovsky studied engineering and had a strong interest in aviation, having learned to fly whilst a student. In 1939 he joined the NKVD, and despite his youth and relative inexperience was appointed head of S & T at the London residency in February 1941, a position he held throughout the war. He found the section he was to command understaffed. 'There were only three of us,' he later recalled, '… but plenty of work.'[10]

One of the most important British technical developments he was able to furnish Moscow Centre with details of was 'Window'.

Throughout 1942 radar-equipped Luftwaffe night-fighters were inflicting catastrophic losses on RAF Bomber Command. Some method had to be devised to counter the night-fighter, or the night bombing offensive against Germany would have to be suspended. The solution Britain's scientists came up with was a method of interfering with enemy radar by scattering aluminium foil strips, codenamed 'Window', which jammed German radars, effectively blinding the enemy's night air defences. So secret was 'Window' that the British were at first reluctant to deploy it over Nazi-controlled territory, fearing that the Germans would simply copy this highly effective anti-radar weapon and use it against them during their own raids on Britain. After much delay and high-level debate, 'Window' was finally cleared for use and employed by the RAF for the first time during Operation Gomorrah, the devastating series of raids on Hamburg, beginning on the night of 24/25 July 1943. It proved an outstanding success, Bomber Command losing only a dozen aircraft out of a raiding force of 791 on that first night.

Although the secrets of 'Window' had been shared with the Americans, who christened it 'chaff', Churchill was not prepared to bring Moscow in on the secret. The Soviets, however, acquired details of 'Window' through one of their British spies, Douglas Springhall.

Born in London in 1901, Springhall had been a stoker in the Royal Navy during the First World War but was dismissed from the service in 1920 for fomenting unrest among his fellow sailors. A founding member of the CPGB, he studied at the International Lenin School in Moscow from 1928–31, which was run by the Comintern. After returning to the UK, he became editor of the pro-Soviet newspaper *The Daily Worker*, served as a political commissar in the British contingent of the International Brigade during the Spanish Civil War, and eventually rose to become the CPGB's National Organizer. While the Nazi-Soviet Pact of 1939 disillusioned many British Party members with the Soviet cause, Springhall kept the faith, loyally following Moscow's orders to support the controversial treaty. Regarded by MI5 as a hardline communist 'motivated by an extreme admiration for Russia', Springhall was said to be 'generally disliked and regarded as a ruthless bully by many of the [CPGB] rank and file.' Understandably judged a security risk, when the Second World War broke out MI5 recommended that he be exempted from military service on the grounds that he 'would certainly constitute a serious menace to morale and discipline.'

Using the cover name 'Peter', in October 1942 he recruited fellow Soviet sympathizer Olive Sheehan, a clerk at the Air Ministry, as a sub-agent. Among the classified material she passed on to Springhall were details of 'Window'. Sheehan also supplied Springhall with the names of other Air Ministry employees she believed shared their pro-communist views. Their activities were exposed when Sheehan's flatmate and Air Ministry co-worker, Norah Bond, became suspicious of her regular meetings with the mysterious 'Peter' at their flat and, after overhearing her tell him about tinfoil 'dropped by our bombers over Germany to muck up Jerry's RDF', alerted her superiors at the Air Ministry.

Sheehan and Springhall were arrested in June 1943, the latter at the CPGB HQ in Covent Garden whilst trying to destroy an incriminating document. His captured diary also revealed the names of other contacts who had been supplying him with information, including an SOE officer, Captain Ormond Uren, who provided details of British policy in the Balkans. MI5 thought it probable Springhall had been engaged in espionage for the Soviets since the beginning of the war and discovered that 'among the information passed to the [Soviet] Embassy were aeroplane plans and figures of tank output.'

MI5 was concerned about the political ramifications and the potential damage to the Anglo-Soviet alliance Springhall's arrest might cause. But Sir Alexander Cadogan, the Permanent Under-Secretary for Foreign Affairs, assured the Security Service that they 'need not in any way consider the danger of international repercussions in this case. If the Soviet Government were in fact involved it had put itself so much in the wrong that it did not have a leg to stand on.'[11]

In exchange for leniency, Sheehan agreed to cooperate with the authorities and testified against Springhall at his trial, held in July 1943 *in camera*, due to the top-secret nature of 'Window'. Springhall was found guilty and sentenced to seven years – of which he served half – while Sheehan received three months hard labour. Expelled from the CPGB, after his release Springhall emigrated to the USSR, where he died in 1953, aged 52.

Springhall's exposure revealed the extent of Soviet espionage being carried out in the UK, despite the two countries' allied status. The case prompted senior MI5 officer Guy Liddell to comment in his diary that 'penetration of the [Armed] Services by the Communist Party is becoming rather serious'.[12]

Besides its network of agents, the Soviets also gained access to the latest British aviation technology from the wrecks of RAF aircraft that crash-landed on Soviet soil. On 11 September 1944 Bomber Command sent a force of 38 Lancasters carrying 12,000lb 'Tallboy' bombs to Yagodnik airfield near Murmansk. From there, they would launch Operation Paravane, an attack on the mighty German battleship *Tirpitz*, anchored in Kaafjord in northern Norway, which had stubbornly resisted multiple British attempts to sink her.

Encountering thick fog at Yagodnik, several of the Lancasters made forced landings on Soviet territory and were damaged. The remaining Lancs carried out the raid on the *Tirpitz* from Yagodnik on 15 September, badly damaging the battleship.[13] Six Lancasters damaged in the forced landings in Russia were left behind and seized by the Soviets. These aircraft were fitted with the latest British technology, including the H2S ground-scanning radar and Gee-H radio-navigation aid, both designed to improve bombing accuracy at night and in poor weather conditions. All of this state-of-the-art equipment, AVM Douglas Colyer pointed out at a meeting of the Chiefs of Staff committee on 27 October 1944, 'would presumably by now have been examined by the Russians.'[14]

Spies in America

Hitler's invasion of the USSR also brought about a fundamental shift in political attitudes in Washington. Although still neutral, the Roosevelt administration was sympathetic to the Soviet plight and in October 1941 extended the Lend-Lease scheme to the Soviet Union. Over the next four years the US would supply vast quantities of materiel, including thousands of aircraft, to the USSR. The NKVD officer Gaik Ovakimian, who supervised S & T collection in the States until his arrest by the FBI in May 1941, was also a beneficiary of the new political climate. In July 1941 the State Department cut a deal with Moscow to exchange him for several US citizens being held in the USSR. Back in Moscow, Ovakimian was put in overall charge of the NKVD's global S & T effort.

With the USSR now on the Allied side in the fight against Nazi Germany, many of the NKVD and GRU's pre-war American spies resumed contact with their handlers, and the Soviet residencies found it much easier to recruit new sources. The Soviets pulled off a major coup by recruiting a spy ring at the heart of Washington's government bureaucracy. Their members included Victor Perlo, a senior statistician at the War Production Board, who furnished Moscow Centre with full details of US aircraft production, and William Ullmann, a Treasury employee who served in the Pentagon during the war. Ullmann gave the Soviets a wealth of data on the USAAF, as well as reports on the testing of the latest aircraft types – although the poor standard of the photographs he took of secret documents meant that much of his material was unreadable.

But Ovakimian, now based in Moscow, was displeased with the work of the agent handlers based at the New York residency, Semyon Semyonov ('Twain'), Nikolai Yershov ('Glan') and F.S. Novikov ('Laurel'). 'Results of the tech. line are unsatisfactory,' he stated in a signal to the chief of the residency, Vasili Zarubin. 'GLAN and

LAUREL don't do anything,' he complained, while 'TWAIN is not functioning at full capacity'. He went on to blame 'inadequate training and management of sources' for the fall-off in the quality of intel being forwarded to the Centre.[15]

Fortunately, Stanislav Shumovsky had by then returned to the US to reactivate his pre-war agent network, with his official cover being that of a member of the Soviet Purchasing Commission, which evaluated and selected aircraft for the Soviet Air Force under the Lend-Lease agreement. One of his reactivated agents was Jones York (aka 'Needle'). Now working for Northrop Aviation, York provided photographs of the blueprints of the company's dedicated night-fighter, the P-61 Black Widow. After moving to Lockheed soon after, he gave the Soviets details of the XP-58 Chain Lightning, a long-range, heavily redeveloped version of their highly successful P-38 Lightning fighter, which in the end never entered production. On 31 October 1943, the Soviet residency in San Francisco reported back to the Centre: '"NEEDLE" has handed over 5 films of material on the "XP-58" and new motors for it.'[16]

After a slow start, Semyonov also confounded critics like Ovakimian, proving himself a highly effective recruiter and handler of agents. In 1943 he was handling no fewer than twenty-eight informants, five of whom specialized in aviation espionage. That year, the New York residency forwarded 211 rolls of microfilm of classified material (compared to 59 the previous year). In 1944, a total of 600 rolls was despatched. In fact, so much classified material was pouring in that the Soviets faced the practical problem of how to send it all back to Moscow. This problem was resolved in November 1942 when an air bridge was established between Alaska and Siberia, allowing the stolen material to be flown to the Soviet Union under the noses of the Americans as 'diplomatic cargo'. With intelligence-gathering in the US back on track, Shumovsky returned to Moscow in May 1943, where he was rewarded for his success with promotion to the TsAGI's Head of Bureau of New Technology.

Now recognized as a top intelligence priority by Moscow, in October 1943 S & T intelligence-gathering underwent a reorganization. It was renamed Line X, and the personnel involved in stealing aviation technology were henceforth known as Line X officers. To handle the upsurge in sources and material from facilities across the States, additional Line X departments were established in the residencies at Washington, Los Angeles and San Francisco.

But not everything ran smoothly. Some of their sources were proving troublesome, notably 'Needle'. In financial difficulties, he demanded more money to continue supplying material. Judging the quality of his intel to be declining, in late 1943 the NKVD decided to drop him as an agent. Even more serious was the feuding within the Soviets' US residencies, which threatened to compromise the entire Soviet intelligence operation in the country. The abrasive manner of the chief resident in New York (and later Washington), the highly experienced intelligence officer Vasili Zarubin, had so offended one of his deputies, Vasili Mironov, that he complained to Moscow Centre about his behaviour, in the hope that he would be recalled. When this approach failed, in August 1943 he wrote an anonymous letter to the head of the FBI himself, J. Edgar Hoover, naming Zarubin and several of his colleagues at the residency – including Semyonov and Leonid Kvasnikov (the deputy resident for S & T intelligence) – as intelligence officers who were involved in stealing US military technology, stating in his letter that they were 'robbing the whole of the war industry of America'.[17] Mironov, said to be suffering from schizophrenia, was recalled to Moscow and sent to a lunatic asylum.[18]

Anxious not to antagonize their Soviet allies, the US Government was slow to act. But in the summer of 1944 Zarubin and Semyonov were finally expelled, while other suspected Soviet spies operating under diplomatic cover were subjected to greater FBI scrutiny. Zarubin's recall did not harm his career. Back in Moscow, he was showered with honours and made deputy head of foreign intelligence of the NKVD due to his success in America.

Despite the disruption, the Soviet intelligence operation continued to function remarkably effectively for the remainder of the war. In 1945, the Soviets sent an astonishing 1,896 rolls of microfilm containing classified material taken by their American agents, a sizeable amount of this intelligence bounty relating to military aviation.

The Soviet spy rings in America were eventually exposed after the war, thanks to a remarkable breakthrough by the US Army Security Agency (precursor of the NSA) in cracking the codes used by the Soviet residencies in America. Codenamed Venona, the programme began in 1943 when the ASA tried to break Soviet codes ahead of the Tehran conference between Roosevelt, Churchill and Stalin to gain an insight into the Soviet leader's intentions. Through great persistence, the ASA codebreakers eventually succeeded in deciphering over 3,000 messages sent from the Soviet residencies to Moscow Centre between 1940 and 1948. Although many of the signals could only be partially recovered, Venona led to the exposure of dozens of American and British citizens who had been – and in some cases still were – spying on behalf of the USSR, among them the atomic bomb spies Klaus Fuchs and Julius Rosenberg, before an NKVD source in the ASA tipped off the Soviets that their codes had been compromised. One of those exposed was Jones York. Arrested by the FBI in 1950, he confessed to his treason but was not prosecuted, as the three-year statute of limitations on criminal prosecution of espionage cases had expired.

The renowned Soviet physicist Abram Ioffe, director of the Academy of Sciences of Leningrad's Physics and Technological Institute, paid handsome tribute to the quality of Soviet S & T intelligence gathered during the war: 'The information [collected] always turns out to be accurate and for the most part very complete . . . I have not encountered a single false finding,' he wrote. The great Soviet aircraft designer Aleksandr Yakovlev also privately acknowledged the debt he owed to S & T in the development of his own aviation projects.[19]

The spoils of war

Besides stealing the secrets of their allies, the Soviets were, of course, also keen to collect as much of the enemy's technology as possible. As German aviation was in many areas far more advanced than the Allies', particularly in the fields of jet propulsion, rocket motors, guided missiles and aerodynamics, this was a priority target for the NKVD and GRU. Efforts to secure examples of the latest German aircraft and engines, as well as recruit the services of the country's leading designers and engineers – willingly or not – began even before Nazi Germany's surrender and continued for several years thereafter.

As Hitler's Third Reich crumbled in the spring of 1945, Britain and the US were also determined to seize as much enemy technology and key personalities of Germany's aviation industry as possible, both for the benefit of their own industries and – even more importantly as East-West relations deteriorated and the seeds of the new Cold War were being sown – to deny these valuable assets to the Soviets.

The main British unit tasked with securing Nazi technical secrets and personnel was No 30 Assault Unit, also known as 'T-Force' (Target Force), which comprised Royal Navy, Army and RAF personnel, with some test pilots from the RAE's Enemy Aircraft Flight attached. The American effort, meanwhile, was codenamed Operation Lusty and was carried out by Air Technical Intelligence teams of the USAAF, nicknamed 'Watson's Whizzers' after their CO Lieutenant Colonel Harold Watson, a former Wright Field test pilot.

Although the Anglo-Americans succeeded in spiriting many of Nazi Germany's leading engineers out of the Soviet-controlled zone of Germany, many others were rounded up by the NKVD and put to work in the Soviet aircraft industry. Some worked willingly for the Soviets, attracted by the high wages on offer, but in other cases coercion was used. To prevent German aeronautical and other scientific engineers being poached by the British, French and Americans, in October 1946 the NKVD's post-war successor, the MVD, implemented Operation *Osoaviakhim*, the transfer of over 2,500 former Nazi engineers in the Soviet zone of occupation to the USSR, of whom around half had worked in the aviation industry.

Among those sent to the Soviet Union were Professor Günther Bock, director of the DVL (*Deutsche Versuchsanstalt für Luftfahrt* – the German Institute for Aviation Research) in Berlin; Siegfried Günter, a senior designer at Heinkel, who had worked on the company's He 219 night-fighter and He 162 *Volksjäger* ('People's fighter') jet interceptor; senior designers at the Junkers' works, Hans Wocke and Brunolf Baade, both of whom had worked on the Ju 287 jet bomber, with its distinctive forward-swept wings; and Ferdinand Brandner, who developed Jumo turbojet engines for Junkers. Lieutenant Colonel Grigori Tokaev, a Soviet aeronautical engineer who helped gather intelligence on Nazi jet aircraft and the V2 rocket as part of the Soviet Control Commission, and who later defected to the British, revealed that relations between the German technicians and the Soviets were often strained, with an atmosphere of mutual suspicion prevailing.

The cutting-edge Nazi designs and technology seized by the Soviets would heavily influence the country's first generation of post-war aircraft, although national prestige demanded that the contribution made by the USSR's erstwhile enemy be downplayed. Ferdinand Brandner, for instance, designed the Kuznetsov NK-12 engine for the Soviets. The most powerful turboprop engine in the world, the NK-12 powered Tupolev's mighty Tu-95 'Bear' strategic bomber, introduced in 1956 and which remains in service with the Russian Air Force's Long-Range Aviation.

Chapter Three

The 'Bull' and 'the Bomb'

On 25 September 1941, amid panic and chaos in Moscow as the Germans advanced relentlessly towards the capital, the Soviets' top spy handler in Britain, Anatoli Gorsky, informed Moscow Centre that he'd received from one of his agents 'a most secret report of the [British] Government Committee on the development of uranium atomic energy to produce explosive material', which had been submitted the day before to the War Cabinet.[1]

This was the first solid intel the Soviets had received of the research being carried out in the West on the practicality of manufacturing an atomic bomb, the initial development of which was being conducted by the British, under the cover name 'Tube Alloys'.

The Committee referred to in Gorsky's message was the Scientific Advisory Committee, and the NKVD source who was providing the Soviets with the information on the group's findings was John Cairncross.

A private secretary to the committee's chairman Lord Hankey, Cairncross was a member of the infamous 'Cambridge Five' spy ring (the others being Kim Philby, Guy Burgess, Donald Maclean and Anthony Blunt), who were recruited by the NKVD in the 1930s.

Recognizing that such a weapon had the potential to fundamentally alter the balance of power in the world, Stalin ordered his intelligence services to make an all-out effort to acquire all the necessary information from the West to allow the USSR to build its own atomic bomb. The huge intelligence operation launched to achieve this went by the appropriate codename 'Enormoz'.

Soviet intel regarding 'Enormoz' was initially collected in Britain, where all the early theoretical work on atomic weapons was being carried out. An early atomic bomb spy – to this day known only as 'K' – provided a number of classified documents on 'Tube Alloys' to the London residency's S & T specialist Vladimir Barkovsky from late 1942. But Moscow's most highly placed informant was Klaus Fuchs, a brilliant German physicist and committed communist who had fled to Britain in 1933 after the Nazis came to power, and went on to become one of the project's leading scientists.

Fuchs and 'K's' information was supplemented later in the war by material passed on by the Soviets' longest-serving British agent, Melita Norwood (codenamed 'Hola'), whose position as a secretary at the British Non-Ferrous Metals Research Association in London allowed her access to many classified documents connected to 'Tube Alloys'. Her material was assessed by Moscow Centre as making 'a valuable contribution to the development of work' in the field of atomic research.[2]

Lacking the money and resources to turn its theoretical scientific work into a practical, viable weapon, the British Government turned over its research to the United States, once the two countries had become formal allies following the Japanese attack on Pearl Harbor. In June 1942 President Roosevelt ordered work to begin on building an atomic bomb, this programme being codenamed the 'Manhattan Project', which came under the direction of the brilliant scientist Robert Oppenheimer.

With the British frozen out of the Manhattan Project until an agreement was reached between Churchill and Roosevelt in July 1943 authorizing full Anglo-American cooperation, the focus of 'Enormoz' switched to the US. The NKVD had little difficulty recruiting informants among the young, idealistic, mostly left-leaning scientists and engineers employed at the Los Alamos facility in New Mexico, where the Bomb was being built. Among their most important atomic spies were Theodore 'Ted' Hall, a Harvard-educated physicist who was just 19 when he began working for the Soviets, US Army sergeant David Greenglass, who worked as a mechanic at Los Alamos, and 'Mar' and 'Kvant' (who remain unidentified to this day). Even Oppenheimer was rumoured to be a Soviet spy, though these claims have never been substantiated. Fuchs, meanwhile, became a key source when he was transferred from London to Los Alamos in August 1944, joining the Theoretical Physics Division.

Igor Kurchatov, scientific chief of the Soviet atomic bomb project, informed NKVD chief Lavrenty Beria in March 1943 that the intel provided by the organization's agents attached to the Manhattan Project 'has made it possible to obtain important guidelines for our own scientific research, bypassing many extremely difficult phases in the development of this problem, learning new scientific and technical routes for its development, establishing three new areas for Soviet physics, and learning about the possibilities for using not only uranium-235 but also uranium-238.'[3]

By the time the Americans dropped the 'Fat Man' and 'Little Boy' atomic bombs on Hiroshima and Nagasaki in August 1945, bringing the war to an abrupt end and ushering in the nuclear age, the Soviets had acquired from their British and American sources almost all the information they required to build their own Bomb and had begun the programme to achieve this objective.

But the Bomb by itself would be of no use without a suitable aircraft to deliver it. The problem facing the Soviets was that they lacked a heavy, long-range strategic bomber capable of carrying such a weapon.

And so, as work progressed at the Arzamas-16 facility in the Nizhegorod region on the development of their copy of the atomic bomb, Stalin directed the Soviet Union's top aircraft designers to solve this problem with all haste.

Red Star 'heavies'

Prior to the Second World War, Imperial Russia and its successor state, the USSR, had a long and illustrious tradition of designing and building strategic heavy bombers. This tradition stretched back to before the First World War, with the *Ilya Muromets* (also known as the S-22), the world's first four-engine heavy bomber. Designed by the legendary Igor Sikorsky and named in honour of a figure from Slavic folklore, the *Ilya Muromets* was derived from a luxury passenger plane and made its first flight on 11 December 1913. Powered by four German Argus engines and with a wingspan of almost 100 feet, it was capable of carrying almost 600kg of bombs and could remain in the air for around five hours.

During the First World War, the *Ilya Muromets* equipped the world's first dedicated heavy bomber squadron, the *Eskadara Vozdushnykh Korablei* (Squadron of Flying Ships), and from February 1915 was used to bomb German positions on the Eastern Front. Of the eighty or so built, only one is thought to have been shot down, although many more were lost in accidents. After the Russian Civil War, several survivors formed the basis of the USSR's first mail plane service.

In 1922 an OKB was founded by an aircraft designer whose name would become synonymous with Soviet bombers, and especially long-range strategic bombers: Andrei Tupolev. The first bomber to emerge from the Tupolev OKB was the ANT-4 (given the military designation TB-1, standing for *Tyazehelyiy Bombardirovshhik* – 'Heavy Bomber'), a twin-engine aircraft which made its maiden flight on 25 November 1925. This was followed by the ANT-6. The world's first strategic heavy bomber of all-metal construction, the ANT-6 was essentially an enlarged, four-engine version of the ANT-4, with increased bombload and range. With test pilot Mikhail Gromov at the controls, it first flew on 22 December 1930 and entered service with the Soviet Air Force, as the TB-3, in 1932. Throughout the 1930s the TB-3 set several distance and payload records, and in the desperate early years of the Great Patriotic War – as the Second World War is known in Russia – it was used as both a bomber (mainly operating at night) and military transport, despite by then being quite obsolete.

An enlarged version of the ANT-6/TB-3, the ANT-16 (TB-4) was subsequently developed but the type was shelved in 1934 before entering full production. Nevertheless, the experience gained on the ANT-16 helped Tupolev and his team in the development of their most ambitious project yet, the ANT-20 'Maxim Gorky' experimental long-range aircraft.

Powered by no fewer than eight M-34FRN engines (six mounted on the wings, with two more in a pusher-puller configuration mounted on struts atop the fuselage), and with a wingspan of 63 metres, the 'Maxim Gorky' was a true giant, at the time the largest aeroplane in the world. Mikhail Gromov piloted the aircraft on its maiden flight on 17 June 1934 and set a world payload record of 15,000kg. Unfortunately, the 'Maxim Gorky' was destroyed in a crash the following year when a Polikarpov I-5 fighter collided with it during a demonstration flight over Moscow, with the loss of 45 lives.

Three more examples were produced, these featuring six engines, as passenger and freight aircraft. Some thought was also given to adapting the ANT-20 into a bomber and troop transport, along with another multi-engined 'super heavy bomber', the ANT-26 (TB-6), which would have a wingspan of 95 metres and be powered by no fewer than twelve engines. Construction of the prototype was well underway when, in July 1934, work on both the ANT-26 and the bomber version of the ANT-20 was halted when it was belatedly recognized that lumbering aircraft of such enormous size would be far too vulnerable to enemy fighter attack.

Although dominant, the Tupolev OKB did not have a monopoly on Soviet heavy bomber designs in the interwar period. The Kalinin

OKB, headed by Konstantin Kalinin, also attempted to meet the air force's requirement for a strategic bomber. The Kalinin K-7 had seven engines and could carry 9,600kg of bombs, over a maximum range of 1,600km. Only one K-7 was built, making its first flight on 11 August 1933. After this prototype was destroyed in a crash three months later, the project was cancelled.

In October 1934 the VVS issued a specification to the Tupolev OKB for a replacement for the TB-3, capable of carrying 4,000kg of bombs over a range of 3,800km. This somewhat ambitious specification was later revised, with a more realistic 2,000kg bombload and range of 3,000km being called for. Valdimir Petlyakov – who had also worked on the ANT-20 and TB-3 – was put in charge of the design work on what was designated the TB-7. The prototype was completed in November 1936 and flew a month later, proving to possess performance similar to that of the American Boeing B-17 'Flying Fortress'. But development of the TB-7 was bedevilled by reliability problems with the engines. The programme was further set back by the Great Terror, which saw both Tupolev and Petlyakov incarcerated in a *sharushka*,[4] and the execution in July 1938 of the C-in-C of the VVS, General Yakov Alksnis, who was a proponent of long-range bombers.

Persistent technical issues, together with a shift in military thinking amongst the VVS high command, which increasingly favoured twin-engined tactical machines like the Petlyakov Pe-2 and Tupolev Tu-2 over large heavy bombers, ensured that only 93 TB-7s would be built. Renamed the Petlyakov Pe-8 in 1942, the survivors of the Nazi onslaught in June 1941 saw limited service in the Great Patriotic War, taking part in the second Soviet air raid on Berlin on the night of 9/10 August 1941 and carrying out sporadic raids thereafter. A Pe-8 was also used to transport Soviet foreign minister Vyacheslav Molotov from Moscow to Britain in 1942 for talks with Churchill.

With only the obsolete TB-3 and trivial numbers of unreliable Pe-8s available to them, the Soviets turned to the US to satisfy their need for a modern heavy bomber. In August 1941 the Soviet Purchasing Commission in the US requested Boeing B-17s. But this, and later appeals for both B-17s and Consolidated B-24 Liberators, were rejected by Washington. Requests to Churchill for heavy bombers were also rebuffed, the only British four-engined type voluntarily handed over to the Soviets being a single Shorts Stirling, which was delivered at Tehran in the summer of 1945.

British and American reluctance to supply the Soviets with 'heavies' was due to the fact that all such aircraft were badly needed for the Anglo-American strategic bombing campaign against Nazi Germany. Another, unspoken, reason for refusal were fears in London and Washington that, given the precarious nature of the West's alliance with Stalin, and Churchill and Roosevelt's distrust of the dictator, such machines could well be used against their countries at a later date.

The Soviets did, however, retain several examples of the B-17 and B-24 which had force-landed on Soviet territory during the course of USAAF bombing missions against Germany and Japan, and incorporated these into the newly established ADD from 1942. They also acquired two Avro Lancasters which had crash-landed in heavy fog near Kegostrov during Operation Paravane (the attack on the *Tirpitz*) in September 1944. Restored to flying condition, they served with transport squadrons of the VVS for a short time before being retired.

As none of these types were suitable carriers of an atomic bomb, in the autumn of 1943 Stalin ordered the Myasishchev and Tupolev OKBs to design a long-range, high-altitude strategic bomber capable of carrying a very heavy bombload. Myasishchev's offering was the DVB-202, which closely resembled the B-29 'Superfortress'. The design was refined in the DVB-203, but this failed to make it off the drawing board. Tupolev's response to Stalin's demand for a strategic bomber was known during its development as 'Aircraft 64' – also designated the Tu-10. The '64' was to feature heavy defensive armament, possess a range of 5,000km and a bombload capacity of 10,000kg. A full-scale mock-up was produced in September 1944, featuring twin tail fins and a distinctive 'bug-eye' cockpit canopy, and production of a prototype was approved a few months later.

Stalin also toyed with the concept of a long-range rocket-powered bomber as a potential carrier for the Soviet atomic bomb. Shortly after the war, Grigori Tokaev, a leading Soviet aeronautical engineer and colonel in the GRU, was tasked with bringing the Austrian rocket scientist Eugen Sänger to Moscow (by kidnapping him, if necessary). Sänger was the designer of the *Silbervogel* ('Silverbird'), an incredibly ambitious Nazi bomber project that was to be powered by rockets and which, he envisaged, would be capable of flying in the upper atmosphere at speeds of up to 8,947mph and could strike targets anywhere in the world. Tokaev understandably considered Sänger's concept – which never left the drawing board – wholly impractical, though according to a contemporary MI5 report, 'Moscow expressed great interest in this project.'[5]

Meanwhile, work on the '64' was proceeding slowly, with Tupolev and his team struggling to overcome various technical issues, mainly associated with the complex electrical systems.

In response to these delays, Stalin decided to opt for a proven aircraft to become the Soviet Union's first atomic bomber – unwittingly provided by his American ally.

Gifts from God

During a trip to the USSR in early July 1943 in his capacity as a special advisor to the US Government on aviation matters, the famous First World War fighter ace Captain Eddie Rickenbacker casually mentioned to his Russian hosts that the US was developing a new 'superbomber'. Soviet intelligence immediately set to work gathering more information on this new aircraft and soon learned that Rickenbacker had been referring to the Boeing B-29 'Superfortress', the first detailed intelligence on which reached Moscow via their spy in the Pentagon, William Ullmann. With a wingspan of 43 metres, a combat range of 2,600km, a pressurized cabin allowing it to operate at altitudes up to 33,000ft, and a maximum bombload of 9,100kg, the B-29 was the most advanced bomber in the world and the single greatest industrial project undertaken by any nation during the Second World War.

The Soviets wasted no time in trying to obtain the B-29. On 19 July 1943 General Belyaev, head of the Soviet Purchasing Commission in the US, made a request for B-29s through the Lend-Lease scheme. This was flatly rejected, as were subsequent appeals. Undeterred, in early 1945 Stalin personally suggested to the US Ambassador in Moscow, Averell Harriman, that the Soviet Air Force could join the USAAF's bombing offensive against Japan if they were willing to provide him with 120 B-29s. Harriman raised practical objections to Stalin's proposal, pointing out that it would take a minimum of six months to train Soviet personnel on the type. Stalin countered that the Red Air Force would select only their most experienced bomber crews for conversion training and would therefore be able to build up its heavy bomber force much more quickly. The US refused to budge.

Fortunately for Stalin, several B-29s fell into his hands after making emergency landings on Soviet territory in the course of carrying out raids on Japanese targets. The first of these (serial 42-6256 and known as 'Ramp Tramp') arrived on 20 July 1944. Heavily damaged by AA fire during a raid on a Japanese steelworks in Manchuria, the B-29's pilot Captain Howard R. Jarnell made an emergency landing at the Tsentralnaya Uglavaya airfield, a few miles east of Vladivostok. The crew destroyed all their documents, radio equipment and the aircraft's Norden bombsight before Soviet troops arrived at the scene. A second B-29, belonging to the 395th Bomber Squadron, crashed in Khabarovsk near the Chinese border after being hit by flak whilst bombing Yawata in Japan a month later, the crew bailing out of the stricken Superfortress. The third example to land on Soviet soil was 42-6365 (named the 'General H.H. Arnold Special', in honour of the commander of the USAAF, Gen Harold 'Hap' Arnold) of the 794th Bomber Squadron, whose pilot Captain Weston Price also made a forced landing at Tsentralnaya Uglavaya, on 11 November. A few days later, on the 21st, a fourth crippled B-29 ('Ding Hao!') was also interned by the Soviets, having made an emergency landing on three engines at Vladivostok.

As the Neutrality Pact signed by Moscow and Tokyo in April 1941 was still in effect, all these aircraft and their crews were interned by the Soviets. While the crews were eventually quietly repatriated, the Soviets held on to their aircraft. Lieutenant Colonel Vasily Reshetnikov, a decorated Soviet bomber pilot who served in ADD during the Second World War, described the arrival of the B-29s in his memoirs as a 'gift from God'.[6]

Impatient at the slow pace of progress on Tupolev's troubled 'Aircraft 64' project, in June 1945 Stalin ordered the designer to concentrate instead on reverse-engineering the B-29, giving him a very tight timeframe of just two years to get his version into the air (Tupolev considered three to be more realistic).

Reverse-engineering American aircraft was not a new concept in the Soviet aviation industry. The Soviets' standard military transport from the early 1940s was the Lisunov Li-2, a locally-built copy of the Douglas DC-3. But Andrei Tupolev, who thought his 'Aircraft 64' design superior to the B-29, was lukewarm about Moscow's policy of reverse-engineering Western aircraft, observing that it 'would take at least two years for us to put foreign aircraft into production in our factories and we would end up with obsolete aircraft instead of modern ones.'[7] But as disagreeing with Stalin was not a healthy option, he dutifully applied himself to the task the Soviet leader had set him. The three airworthy B-29s were initially ferried to Izmaïlovo airfield on the outskirts of Moscow. Of these three, the 'General H.H. Arnold Special' was chosen as the main template for the Soviet version of the

B-29, being the least damaged, and this was sent to Khodynka airfield in Moscow, with 'Ding Hao!' used for reference purposes. Meanwhile, the first B-29 to land on Soviet soil, 'Ramp Tramp', went to the LII to serve as a training platform for Soviet crews.

The LII (Flight Research Institute) was established in the town of Ramenskoye, 40km southeast of Moscow, on 8 March 1941, under the command of the famed test pilot Mikhail Gromov. The Institute's principal role was to conduct flight tests of prototype and experimental Soviet aircraft, and all the major OKBs had facilities there. After the Second World War, captured Western aircraft and aviation equipment were also tested and evaluated at Ramenskoye. One of the largest air bases in Russia, it has three runways, one of which – at 5,403m – is the longest in Europe.[8]

Pirating the B-29 was an enormous undertaking, the largest and most complex project embarked on by the Soviet aviation industry up to that time, and one which would eventually involve more than 900 organizations. Soviet technology and manufacturing methods lagged well behind those of the United States and would have to be thoroughly modernized if the project was to become a reality. Another major challenge confronting Tupolev's team was the need to convert US imperial measurements to their closest metric equivalents.

Initially given the designation B-4, much of the work on the bomber was carried out at the GAZ-22 works at Kazan. The greatest divergence with the American original was the substitution of the Curtiss-Wright R-3350 Duplex-Cyclone engines for Shvetsov ASh-73TK units, although the Cyclone's turbo-chargers were copied. The Superfortress's defensive armament of Browning .50 calibre machine guns also gave way to more powerful 23mm NR-23 cannon.

After overcoming many technical challenges, the first Tu-4 (as the B-4 had been renamed) was completed in February 1947 and made its maiden flight on 19 May, thereby – just – meeting Stalin's two-year deadline from initiation of the project to flyable aircraft. The first public appearance of the Tu-4 was at the Tushino Air Display on 3 August 1947. Western military observers attending the air show initially assumed the three examples that took part were the original B-29s interned during the war. But they soon realized that the Soviets had succeeded in making their own version of the Superfortress, which was assigned the NATO reporting name 'Bull'. By the time production ended in 1952, around 850 had been built. Ten of these were modified for the nuclear role and designated the Tu-4A (the 'A' standing for *Atomniy* – 'atomic').

The Soviets had succeeded in producing their first atomic bomb, the RDS-1 (broadly similar to the American 'Fat Man' device) in 1949. Codenamed 'First Lightning' (and dubbed 'Joe-1' by the Americans), the 22 kiloton device was successfully detonated in a static ground test at Semipalatinsk in Kazakhstan on 29 August 1949. This event can be considered the true start point of the Cold War. Although practice drops from Tu-4s had been performed using dummy nuclear bombs since 1948, it wasn't until 1951 that the first test of a live atomic bomb dropped from the air took place.

On 18 October, a Tu 4A flown by Lieutenant Colonel Konstantin Oorzhuntsev – who had been made a Hero of the Soviet Union during the war – and carrying an improved RDS-3 model atomic bomb with a yield of 41 kilotons, took off from Zhana-Semey airfield in Kazakhstan. A few minutes before 10 am the crew jettisoned their bomb, from an altitude of 32,800ft. The explosion and its immediate aftermath were later vividly described by the aircraft's bomb-aimer, Captain Boris Davydov:

The first thing we knew was a tremendous flash. Then the first and quite powerful blast wave caught up with the aircraft, followed by a weaker second wave and a still weaker third wave. The flight instruments went crazy, the needles spinning round and around. Dust filled the cabin, even though the aircraft had been vacuumed clean before the flight. Then we watched as the dust and debris cloud grew; it quickly mounted right up to our own flight level and billowed out into a mushroom. There was every thinkable colour and shade to that cloud. I am lost for words to describe what I felt after I had dropped the bomb.[9]

Thanks in large part to its spies in Britain and the US, the Soviet Union had now become the world's second nuclear power. What's more, again thanks partly to its spies, the USSR now also possessed a delivery platform capable of striking anywhere in Western Europe from Soviet territory.

Chapter Four

Project *Vozdukh*

Regarded as the father of the jet engine, Frank Whittle, a 22-year-old Flying Officer in the RAF, patented the gas-turbine engine in 1930. Unable to attract the interest of the Air Ministry for his revolutionary invention, Whittle was forced to pursue his concept as a private venture, and it wasn't until May 1941 that Britain's first jet-powered aircraft, the Gloster E.28/39, took to the air. Official inertia allowed Nazi Germany to steal a march on the British, and on 27 August 1939 the world's first jet aircraft, the Heinkel He 178, made its maiden flight, powered by an HeS 3 turbojet engine developed by the German engineer Hans von Ohain.

Concurrent with these developments in Europe, research into jet technology was also ongoing in the USSR. Much theoretical work on jet propulsion had been carried out by Boris Stechkin in the pre-WW2 period. In 1939, the Ukrainian-born engineer Arkhip Lyulka produced designs for the Soviet Union's first gas-turbine engine at the Kharkov Aviation Institute. He continued his research at a design bureau in Leningrad, but when Nazi forces laid siege to the city from September 1941, Lyulka and his team were evacuated to the TsIAM in Moscow to continue their work.

At the same time, the Soviets were also experimenting with ramjet fighters. A simplified type of jet engine with no turbines, a ramjet boosts the performance of an aircraft for short periods by using its forward momentum to compress air, which is then ignited in a combustion chamber. The first aircraft to utilize this power system was a modified Polikarpov I-153 biplane, which flew in October 1940.

Later, a Yak-7B fighter, designated the Yak-7 PRVD, was tested with a pair of improved DM-4S ramjets. Flight testing began in May 1944, but the ramjets added considerable weight and drag to the airframe, resulting in only a marginal improvement in top speed. Fuel consumption was also excessive, and therefore no fighters powered by this system entered production.

Work on Lyulka's gas-turbine engine, meanwhile, was progressing at a slow pace. To speed up development of a Soviet jet engine, in 1943 Moscow Centre instigated Project *Vozdukh* ('Air'), an intelligence-collection effort aimed at acquiring the secrets of Britain and America's jet engine programmes. This was one area of technical development neither Churchill nor Roosevelt was willing to share with their Soviet ally. As Churchill told the Chiefs of Staff in 1942, 'there were certain items of equipment which it would be in the common interest of the Allies to keep secret from each other.'[1]

Although Moscow Centre had learned from a source in the US aviation industry that the British were well ahead of the Americans in jet technology, a signal from the New York residency informing them that 'development of [jet] aircraft construction in the ISLAND [codename for the UK] … outstripped the COUNTRY [the US]', the Soviets found it easier to recruit engineers and designers involved in jet engine development in America.[2]

In charge of Project *Vozdukh* was Andrei Shevchenko (codename 'Arsenij'). An NKVD officer and qualified aeronautical engineer, Shevchenko arrived at the New York residency in summer 1942 to head up the Soviet Purchasing Commission's aviation department.

Much of the work on jet propulsion in the US was being carried out by the Bell Aircraft Corporation, which produced America's first jet aircraft, the P-59 Airacomet, in October 1942. Conveniently, Bell was supplying large numbers of its piston-engined P-39 Airacobra and P-63 Kingcobra fighters to the USSR under the Lend-Lease scheme, and Shevchenko was attached to the company as the official Soviet liaison.

In 1943 he recruited Leona Franey, a secretary at the company who was in charge of its vast archive of manuals and technical documents, as a source. Shevchenko, whom she later said was 'always very, very nervous' during their meetings, persuaded her to photograph documents relating to the P-59 and the company's research on swept-wing designs, paying her $30 for each photographed document. In late 1944 Shevchenko recruited another important source at Bell, Larry Haas, an engineer whose duties included training Soviet technicians sent to Bell's factory in Niagara Falls, Buffalo on the P-39 and P-63 fighter. A November 1944 signal sent from the New York residency to Moscow Centre disclosed that Arsenij had met with Haas in the latter's office, where he 'produced from the safe a detailed drawing of the I-16 and gave a full description and particulars.' The General Electric I-16 (later redesignated the J31) was America's first mass produced jet engine, which powered the Bell P-59.

In early 1945, Haas left Bell to join the Westinghouse Electric Company at its facility in Philadelphia, where he worked on the development of gas-turbine engines. At a hearing of the House Un-American Activities Committee in 1949, which was investigating Soviet penetration of American industry and politics, Haas revealed: 'At that time this organization was perhaps the furthest advanced in gas-turbine engines, since they were the forerunners in development of what is referred to as the axial-flow type of compressor, and this type has been the forerunner of today's most powerful engines. So in reality that was a most opportune position for anyone working with the Russians to be in.' He also testified at the hearing that in March 1945 Shevchenko asked him to provide 'highly secret turbojet engine data, drawings, reports, and other pertinent information.' Between then and November 1945, Haas admitted that 'innumerable exchanges of microfilm were made with Mr Shevchenko.'

Unknown to Shevchenko, however, Franey and Haas were working for the FBI as double agents, and the documents they supplied were either unclassified or false. 'The material was all screened by the Bureau of Aeronautics and the Federal Bureau of Investigation before it was handed to him,' confirmed Haas. 'The data which we gave Mr Shevchenko while I was in Philadelphia was of such a nature as to be greatly misconstrued, misleading, and entirely wrong.'[3]

But there were other genuine informants in America spying for the Soviets who remained undetected. The most important of these aviation spies was William Perl (also known as William Mutterperl). Born in 1919, Perl studied engineering at the City College of New York, during which time he became involved in communist youth politics and joined the spy ring headed by the infamous Julius Rosenberg, who would go on to serve as the main contact for the atomic bomb spies. After graduating, in 1940 Perl landed a job at NACA, at the Langley USAAC air base. Four years later, he transferred to NACA's Lewis Flight Propulsion Laboratory in Cleveland. This gave him access to blueprints of the latest aircraft and engine designs, including gas-turbine powerplants, from America's leading aviation companies.

Described by one of his handlers, Aleksandr Feklisov, as 'a well-built, handsome fellow, over six feet tall, and a smart dresser,' Perl was initially given the rather inappropriate codename 'Gnome', which was later changed to 'Yakov'.[4] Amongst the intel he provided to the Soviets was a complete manual of the Lockheed P-80 Shooting Star, the USAAF's first operational jet fighter, and details of the unorthodox Consolidated Vultee XP-81. A signal dated 18 May 1944 forwarding information supplied by 'Yakov' on the latter was intercepted by the US Army's Security Agency and later partially decoded as part of the Venona programme. The signal disclosed that the XP-81 was 'a long-distance fighter . . . intended to escort bombers at great altitude,' for which it was 'provided with two power installations.' To resolve the major drawback of the early jet aircraft – their short range due to heavy fuel consumption – the XP-81 was powered by a combination of a General Electric J33 turbojet, fitted in the rear of the fuselage, and a General Electric TG-100 turboprop engine in the nose, driving a propeller, for cruising work. 'With the aid of full power from both assemblies,' the signal revealed, the XP-81 was expected to attain 'a speed of 500 miles per hour.'[5] A Consolidated Vultee press release boasted: 'Because the two engines operate separately, the fighter can fly on either one or both. Their combined horsepower is virtually the same as that produced by all four engines on a heavy bomber.'[6]

The first of two prototypes that were built made its maiden flight at Muroc air base in California on 11 February 1945. But by that late stage in the war, the USAAF's need for a long-range escort fighter was dwindling, and the project was eventually cancelled in 1947.

Perl's reports on the XP-81 may have influenced the Soviet decision to commission its own mixed-powerplant fighters, the MiG I-250 and the Sukhoi Su-5. The I-250 – also known as the MiG-13 – employed a hybrid power system: the Kholschevnikov VRDK (Jet-Propelled Auxiliary Compressor), also known as the Kholschevnikov accelerator, in which a conventional Klimov VK-107 piston engine drove an axial flow compressor positioned in the deep fuselage, providing additional thrust for short periods. On 3 March 1945, the prototype made its first flight, reaching a top speed of 511 mph. The similar Sukhoi Su-5 first flew a month later. Only a single Su-5 prototype and a very small number of I-250s were manufactured, Soviet interest in mixed-powerplant fighters waning as they chose to focus instead on pure jets.

Thanks to information provided by Perl and others, in August 1944 Arsenij was also able to forward to Moscow important technical details on the latest improvements the Americans were making to their jet engines, including a redesign of the combustion chamber to prevent excessive heat build-up – a problem also encountered by the Soviets, and which would bedevil their early jet engines – and simplifying the ignition system.[7]

Perl was considered to be one of Moscow's most important assets in the US during the war years. Feklisov later wrote that 'Yakov' had supplied his handlers with 5,000 pages of classified documents, half of which Moscow Centre assessed as 'very valuable', the other half as 'valuable'.[8] His importance can be gauged by the fact that, when he moved to the Lewis Flight Propulsion Laboratory, the Soviets instructed members of the Rosenberg spy ring to maintain a safe house in Cleveland exclusively to receive and forward Perl's intel.

Although Moscow was receiving a regular flow of high-grade intelligence from across the Atlantic, tensions were surfacing at the New York residency. In the summer of 1944, NKVD officer Stepan Apresyan (codename 'May') arrived to take up the post of acting head of the residency, replacing the veteran spy Vasili Zarubin, who was recalled to Moscow after his cover was blown by an embittered subordinate. His appointment angered Zarubin's deputy Roland Abbiate, who coveted the job for himself and judged Apresyan – who had never served abroad before – too young (he was only 28) and inexperienced for the position. In a withering assessment, he wrote of his new superior: 'MAY is utterly without the knack of dealing with people, showing himself excessively abrupt and inclined to nag . . . A worker who has no experience of work abroad cannot cope on his own with the work of directing the TYRE office [codename for the New York residency].'[9]

More serious than these personality clashes was the decision of Elizabeth Bentley, an important courier for the Soviets' agent network in the US, to turn FBI informer in 1946 following the death by heart attack of her handler and lover Jacob Golos. The information she provided allowed the FBI to dismantle the Rosenberg spy ring. While Perl initially escaped detection, the destruction of Rosenberg's network effectively brought his secret career as a Soviet spy to an end. Thanks to clues obtained through the Venona decrypts, he was eventually identified as a possible Soviet spy by the FBI. Though never prosecuted for espionage, in 1953 Perl was convicted of perjury for giving false testimony at the trial of Julius Rosenberg and sentenced to five years.

By the time the Second World War had ended, little was known in London and Washington about the state of jet engine development in the USSR, as a British intelligence report of September 1946 frankly admitted: 'Some research and development on gas turbines was undertaken by Soviet engineers during the war but little is known about this activity. To judge by the lack of success achieved in the building of turbo-superchargers it is unlikely that Soviet jet engine design is very advanced.'

The exploitation of captured Nazi technology and engineers would, however, greatly speed up development in this field, the intelligence analysts predicted, with the Soviets' main interest lying in 'improving the high-altitude performance and thrust of German gas turbines.'[10]

Amongst the war booty that fell into Soviet hands in 1945 were examples of the Arado Ar 234 bomber/reconnaissance aircraft, and the Heinkel He 162 and Messerschmitt Me 262 jet fighters. On 15 August 1945, Lieutenant Colonel Andrei Kochetkov of the NII Flight Test Section became the first Soviet pilot to fly a turbojet-powered aircraft in Russian airspace when he made a test flight from Shchyolkovo airfield, near Moscow, in a captured Me 262.

The Me 262's powerplant, the Junkers Jumo 004B axial-flow jet engine, was copied and produced in the USSR as the Klimov RD-10. The RD-10 and the RD-20 – a copy of the BMW 003A engine, used in the Heinkel He 162 – became the first gas-turbine engines to be mass produced in the Soviet Union and were used to power the country's first true jet fighters: the MiG-9 and Yak-15.

The MiG-9 was developed in response to an order issued in February 1945 by the Council of People's Commissars for a single-seat, jet-propelled fighter, to be powered by two RD-20 engines. The Mikoyan OKB played it safe with a conventional, straight-wing design, with the two RD-20 units housed in a barrel-shaped fuselage. The first of three prototypes, designated the I-300, was completed in early

1946. At the same time, Yakovlev was developing a rival jet fighter, the Yak-15, which was essentially a standard Yak-3 that substituted its VK-107 piston engine with an RD-10 in a redesigned nose. According to legend, a coin toss decided which of the two would make the first flight. The MiG won, and at Ramenskoye airfield on 24 April 1946 test pilot Alexei Grinchik made history by taking the USSR's first pure jet aircraft into the air.

Both these fighters were considered interim types, intended to help transition pilots from piston to jet power pending the introduction of more advanced designs, and so production was fairly limited, with around 610 MiG-9s and 280 Yak-15s built.

Besides interceptors, the RD-10 was also installed in the EF 131, an early Soviet attempt to produce a jet bomber. A direct copy of Junkers' experimental late-war Ju 287, which featured a highly distinctive forward-swept wing configuration, the EF 131 was intended to form the basis of the Soviets' strategic jet bomber force. One prototype was built, but by the time flight tests began in May 1947, the design was out of date, leading to the cancellation of the EF 131 and its successor, the EF 140, in 1948.

By 1946, Soviet efforts to develop a reliable, high performance jet engine had reached an impasse; in this period of rapid technological progress, the German BMW 003 and Jumo 004 had become obsolescent, while their own designs were proving a disappointment. The Lyulka TR-1, the USSR's first wholly indigenous gas-turbine engine, first ran on 9 August 1946. However, the TR-1 had a very high rate of fuel consumption and, with less than 3,000lbs of static thrust, failed to produce the power demanded of it. Consequently, it was not ordered into large-scale production.

Fortunately for the Soviets, their efforts to build a jet engine with world-class performance were about to receive a major boost from a most unexpected source.

'What fools would sell us their secrets?'

In April 1946, Stalin chaired a meeting of the Soviet Union's top aeronautical engineers at the Kremlin to discuss the current state of jet engine development in the USSR. What he heard greatly displeased him.

Given the problems they were experiencing with the Lyulka TR-1, one of those present lightly suggested that they should make an approach to the British to ask if they would be willing to sell them examples of their latest jet engines, the Rolls-Royce Nene and Derwent V, which Soviet engineers had learned about through the British aviation press. This suggestion allegedly prompted Stalin to remark scornfully: 'What fools would sell us their secrets?'[11]

A centrifugal compressor turbojet, the Nene was developed by Rolls-Royce in 1944 as a successor to its Welland engine, which powered the early marks of the Gloster Meteor, the RAF's first operational jet fighter. At the time, it was the world's most powerful turbojet, boasting 5,000lbs of static thrust. The Nene's qualities were also recognised by the Americans, who produced their own licence-built version as the Pratt & Whitney J42. The Derwent V, which entered production in 1945, was essentially a scaled-down version of the Nene, designed to fit the Meteor F.4. On 7 November 1945, a Meteor F.4 powered by Derwent Vs set a new world speed record of 606 mph.

Despite Stalin's scepticism, and no doubt to the general amazement of the Soviets, when officials from their trade office in London approached Rolls-Royce a few weeks later with a view to purchasing the engines, they were met with a positive response. In June 1946, a deal for ten Nenes and the same number of Derwent Vs, at a unit cost of, respectively, £7,300 and £6,050, was reached.

Politics then intervened. Britain's Labour government was divided on whether or not to approve the sale. Foreign Secretary Ernest Bevin strongly opposed the deal, backed by the Air Ministry and the Chiefs of Staff. Sir Stafford Cripps, president of the Board of Trade and a known Soviet sympathizer, on the other hand, argued that the deal was important for the UK's foreign trade. He also raised concerns that rebuffing Moscow could endanger a vital trade deal for Soviet timber and grain imports then under negotiation.

Anxious to maintain a working relationship with Stalin, Prime Minister Clement Attlee was eventually persuaded by the Cripps faction, and in September 1946 he gave the deal the go-ahead. 'I can see no reason for withholding [the engines] from the USSR, whereas their refusal would only cause trouble and suspicion,' he wrote.[12]

The following month, a four-strong Soviet trade delegation arrived at a Rolls-Royce factory in the UK for an initial five-day induction course. A month later, another Soviet delegation arrived, which included the famous engine designer Vladimir Klimov, co-founder of the MiG OKB Artem Mikoyan, and the metallurgist S.T. Kishkin, for a tour of Rolls- Royce's Derby factory. One area of manufacture the British were not prepared to share with the Soviets was the Nimonic alloys used for the blades of the gas turbines. These alloys were of particular interest to the Soviets, being far superior to those used in their own engines. Klimov, however, cleverly circumvented this restriction by treading through tiny pieces of Nimonic alloy lying on the factory floor, embedding them in his soles and later sending these samples back home.

The trade delegation also tried to negotiate a deal for de Havilland Vampires, which were fitted with the brand new Goblin jet engine. But the tiny size of the order (the Soviets only wanted three aircraft) raised suspicions that they wanted the jets only for technical analysis, and on this occasion Attlee vetoed the proposed deal.

The initial order for 20 Nene and Derwent engines was despatched to the USSR in March 1947. A follow-up order for a further 15 Nenes (including five of the improved Mk II model) and 20 Derwents was subsequently approved and swiftly despatched. The terms of the contract also allowed for up to 20 Soviet technicians to be sent to the

Rolls-Royce factory 'to learn the assembly, operation, maintenance and repair of the engines.'[13] Seventeen engineers duly arrived at the Derby factory in late 1947 for instruction on the engines by their British counterparts.

Back in Russia, the Soviets wasted no time reverse-engineering the Rolls-Royce powerplants at the GAZ-45 and 500 factories. Before the engines had even arrived on Soviet soil, preliminary work on the reverse-engineering process, including producing detailed blueprints of the engines, had already been carried out at Klimov's design office in Leningrad. The Soviet copy of the Derwent V, the RD-500 ('RD' standing for *Reaktivnyy Dvigatel,* meaning jet engine) was completed first and underwent initial testing in December 1947, while the Nene copy (renamed the RD-45) followed soon after.

In breach of the terms of the export deal, the Soviets soon began mass production of their own unlicensed versions of the engines. When they found out, the speed with which the Soviets accomplished this impressed the British Air Ministry: 'Within two years, the copies were coming off the production lines. This was a remarkable achievement,' marvelled a secret intelligence summary.[14] Besides experimental prototypes like the Yakovlev Yak-25, the only production aircraft to make use of the RD-500 were the La-15 and Yak-23. The RD-45 enjoyed much greater success, serving as the powerplant of the Ilyushin Il-28 'Beagle' tactical twin-jet bomber (of which more than 6,000 were built, serving in the air forces of 20 countries) and the first prototypes of the MiG-15.

An improved, more powerful version of the RD-45 was quickly developed, the Klimov VK-1. The VK-1 increased the thrust by twenty per cent and was used in the improved MiG-15bis, which entered production in 1950 and brought the fighter's maximum speed up to 670 mph. 'It is a redesign based on the original Nene, the major differences being in the compressor and turbine so that they could handle the higher mass air flow. The redesign of the outer compressor casing, so that it could handle more air and yet maintain the same overall diameter, is an outstanding example of engineering ingenuity,' the secret British intelligence summary reported.[15] The VK-1 design was also sold to China, where it was produced as the WP-5, powering the Chinese version of the MiG-17, the Shenyang J-5.

The MiG-15's dramatic entry into the Korean War in late 1950, and the realization that the new Soviet jet fighter was powered by a copy of a British engine, ignited a political controversy on both sides of the Atlantic over the UK government's decision to sell the Nene and Derwent to the Soviet Union. US General Arthur Trudeau, who commanded the 7th Infantry Division in the Korean War, asserted that 'if the British had not sold the Nene Rolls-Royce jet engines [to the USSR], there never would have been a MiG in the skies over Korea.'[16]

Criticism was also levelled at Attlee's government in the British Parliament. Conservative MP and former SOE agent Fitzroy Maclean called the handover of British jet technology 'a blunder of the first order.' Labour government minister George Strauss defended the deal, pointing out that 'these engines were no longer on the secret list when supplied to the Russians.'[17]

With the controversy refusing to abate, when he was returned to power in October 1951, Winston Churchill ordered the Ministry of Supply to conduct a thorough investigation into the deal. Allowing Soviet engineers to inspect the Rolls-Royce factories and receive British training on the engines was a particularly contentious aspect of the Anglo-Soviet agreement. However, the Ministry of Supply's report advised Churchill that, under the terms of the deal, the Soviets could have 'charged Rolls-Royce with breach of contract if we had not permitted the engineers to attend the course.' Attlee's government had also received assurances that 'the school where the Russians were to attend instruction on the assembly, operation, maintenance and repair of Nene and Derwent engines was quite separate from the main

Rolls-Royce works, and the firm said that security measures would be taken to prevent the engineers from obtaining information other than that which they would learn during the course.' The report added that the Soviet visitors 'were not allowed to see any secret work in progress.'[62]

Nonetheless, on 27 April 1953, government minister Duncan Sandys stated in Parliament that, in his view, 'the sale of these engines undoubtedly had the effect of reducing by about two years the technical lead in the development of these engines which we then had over the Russians.'[63]

Chapter 5

The Million Dollar Baby

November 8th 1950 was a milestone in the history of military aviation. On that day, the first jet versus jet combat took place. The antagonists were a USAF Lockheed F-80 Shooting Star and a MiG-15 of the North Korean People's Air Force. In response to a plea from the North Korean leader Kim Il Sung for modern fighter aircraft to defend his country from American air raids, in October 1950 Stalin agreed to send the new MiG-15 to the Korean peninsula, along with a cadre of highly experienced Soviet pilots. To disguise direct Soviet involvement in the war, the pilots would masquerade as Koreans and their aircraft carry North Korean markings.

On 8 November four F-80s of the USAF's 16th Fighter Squadron were flying over the Yalu River – an area soon to become infamous to American pilots as 'MiG Alley' – when they spotted around a dozen MiG-15s flying below them. The F-80s descended on the communist jets and in the ensuing dogfight Lieutenant Russell Brown claimed one MiG shot down.[1] Although the Americans were apparently victorious on this occasion, the arrival of the MiG-15 was a dramatic and unpleasant development in the air war for the UN forces, which swiftly brought the air supremacy they had established in the first weeks of the war over the Korean peninsula to an end. The British Gloster Meteor, operated by the Royal Australian Air Force's No 77 Squadron in Korea from July 1951, found itself at a particular disadvantage. The squadron's CO, Wing Commander Gordon Steege, stated in a report that the ageing Meteor was 'vastly inferior in performance' to the Soviet jet, and commented gloomily that 'unless the MiGs are operated unintelligently, Meteors are not going to account for many.'[2]

The threat to UN air operations over North Korea was underscored a few months after that first clash between the Americans and Soviets when, on 12 April 1951, around 30 MiG-15s tore into a formation of 48 B-29s carrying out an attack on a bridge between Korea and China. The bombers' heavy fighter escort comprising around 100 F-80 and Republic F-84 Thunderjets was unable to prevent three of the Superfortresses being shot down and a further seven severely damaged. The MiG-15s suffered no losses. The USAF called this day 'Black Thursday'. Bombing raids over North Korea were suspended for three months in the wake of this attack, and when they resumed were restricted to the hours of darkness.

The Americans' immediate riposte to the MiG scourge was to rush its newest jet fighter, the North American F-86 Sabre, to Korea, to steadily supplant the F-80s and F-84s that were so clearly outclassed. The first F-86s arrived at Kimpo air base near Seoul in mid-December 1950, with the first combat between the two types occurring a few days later. The Americans claimed that by the end of the war the Sabres had shot down a total of 792 MiG-15s, while losing only 78 in return.[3] While this would suggest the F-86 was dominant, today these figures are widely regarded as grossly exaggerated and that the kill to loss ratio was actually much narrower.

Whatever the true figure, the MiG-15 was clearly a formidable adversary, one that shook the West out of its complacency regarding the effectiveness of Soviet warplanes.

Project S

Recognising the limitations of the MiG-9 and Yak-15, in March 1946 the Mikoyan, Lavochkin and Yakovlev OKBs were instructed by the Kremlin to develop a second-generation jet interceptor, capable of achieving Mach 0.9 (667 mph) in level flight. To achieve this, the Mikoyan OKB realized that the new fighter would require a swept-wing design and a powerplant more advanced than the gas-turbine engines currently under development in the Soviet Union. As we have seen in the previous chapter, the delivery of the British Rolls-Royce Nene solved the latter problem.

Given the in-house designation 'Project S' (the 'S' standing for *strelovidnostii*, meaning swept), it had a Nene I engine positioned in the fuselage aft of the cockpit, wings swept back at an angle of 35 degrees, and carried a hard-hitting armament of a single 37mm NS-37 and a pair of 23mm NS-23 cannon. The first prototype of what was then known as the MiG I-310 was completed in November 1947, and the first flight took place from Ramenskoye airfield on 30 December, with Victor Yuganov at the controls. The prototype of Lavochkin's rival project, the La-15, had its first flight just nine days later. Yakovlev's offering, the Yak-30, lagged far behind, its maiden flight not taking place until 4 September 1948.

MiG's I-310 quickly emerged as the clear favourite, and was ordered into production in March 1948 – even before the first flight of the rival Yak-30. As a back-up, the Lavochkin La-15 was put into very limited production, with around 235 being built. The first production examples of what was now designated the MiG-15 began entering service with the VVS from January 1949. Standard production version was the MiG-15bis, powered by the improved Klimov VK-1 engine and incorporating various other refinements, which began entering squadron service in early 1950. It was this version that was sent to Korea by Stalin. An estimated 15,000 MiG-15s of all variants were built before production ended, and it served with over 40 air forces.

To steal a MiG

By 1951, with the MiG menace in Korea growing, the USAF decided that they must acquire as much intelligence as possible on the fighter to unlock its secrets and so devise tactics to combat the new jet. Leading the effort in Korea to achieve this was the controversial intelligence officer Captain Donald Nichols.

Described by one of his superiors as the finest intelligence operative in Korea, Nichols had served as a mechanic in the US Army during the Second World War and transferred to the USAF when it became a separate branch of the military in 1947, specializing in intelligence work.[4] In February 1951 Captain Nichols was serving in Korea, where he formed and commanded a shadowy intelligence outfit within the USAF, known as the Special Activities Unit #1 (later renamed the 6004th Air Intelligence Service Squadron). Made up of both US and South Korean personnel, its tasks included acquiring communist military hardware. When reports reached Nichols that a MiG-15 had crash-landed near the Chongchon River, tantalisingly close to Paengyong-do, an island which was under UN control, he organized a daring mission to visit the crash-site.

On 17 April, Captain Nichols and a team of officers from the 6004th took off from the island's airfield in a US Navy Sikorsky H-19 helicopter and headed to the crash-site, some 80 miles behind enemy lines and close to a major supply depot. Over hostile territory, the H-19 came under small arms fire but successfully landed near the crashed MiG. Nichols was then said to have 'coolly and efficiently photographed the material, recorded all inscriptions and technical data, and supervised dismantlement of vital parts', including the turbine blades and horizontal stabilizer. He and his team then took off again in the now overloaded helicopter. The H-19 again came under ground fire on the return flight, one of the rotor blades being hit, though the pilot managed to land safely back at Paengyong-do.

Nichols received the Distinguished Service Cross for his daring exploit, the citation stating that he 'voluntarily risked his life to wrest information of inestimable value from the very grasp of the enemy.'[5] However, a South Korean member of his team later disputed the official version of events, claiming that Nichols greatly inflated his own role in the operation and didn't even set foot on North Korean soil during the salvage operation.[6]

Whatever the truth, there's no doubt that it was a significant intelligence coup. The recovered parts were shipped back to Wright-Patterson AFB in Dayton, Ohio for analysis by personnel of the Air Technical Intelligence Center where, according to a US intelligence summary, 'a great deal of information was gained.'

ATIC's origins went back to 1917, when the US Army Air Service's Engineering Division established the Foreign Data Section at McCook Field in Dayton, whose job was to study foreign aircraft. In 1945 this evolved into T-2 Intelligence Directorate of the USAAF's Air Materiel Command, relocated to Freeman Field, Indiana, and Wright-Patterson, which evaluated captured German and Japanese aircraft. This in turn became the Air Technical Intelligence Center on 21 May 1951, headquartered at Wright-Patterson and now tasked with analyzing Soviet bloc aircraft, technical documents and aviation-related hardware that fell into American hands. ATIC's commanding officer was Major General Harold E. Watson. He was well qualified for the job, having led a team of aviation experts to collect Nazi aircraft at the end of the war on behalf of T-2.

ATIC personnel who examined the MiG components found that 'general workmanship reflected in all of the turbo-jet engine parts is excellent,' and noted that 'proficiency in the art of welding was particularly evident.'[7]

A few weeks later, Nichols organized a second expedition to visit the MiG-15 crash-site to retrieve more parts from the wreckage. This time, however, the mission ended in tragic failure. His plan was to parachute a 15-strong team of South Korean agents attached to his unit close to the crash-site, where they were to recover the components and transport them by oxcart to the coast, steal a boat, and then sail to nearby Chodo Island, which was occupied by UN forces.

Unfortunately, all but one of the South Korean team were captured by Chinese soldiers immediately upon landing and never heard of again.[8]

'A very valuable prize'

Despite this setback, efforts to obtain a MiG-15 continued. The next opportunity came in July 1951. On the 9th, F–86s of the 4th Fighter Interceptor wing clashed with MiG-15s, one of which was shot down and seen to belly-land in shallow water off the west coast of North Korea. Four days later, reconnaissance photographs taken by Royal Navy Sea Fury fighters from the carrier HMS *Glory* confirmed that the crashed MiG was lying on mudflats 40 miles north of Chodo Island, well inside North Korean territory. Analysis of the imagery showed that, other than missing its tail section, the MiG appeared largely intact, and so it was decided that an operation should be mounted, 'to which the greatest secrecy and importance were attached', to snatch the aircraft from right under the noses of the North Koreans. The plan to salvage the aircraft was named, simply, Operation MIG.

It was to be a multinational effort, involving vessels and personnel of the British, US and South Korean navies, under the overall command of Rear Admiral Alan Scott-Moncrieff DSO, second-in-command of the Royal Navy's Far East Fleet. The target aircraft was codenamed 'Crown Jewels' by the British, 'because it was rather like an attempt to find King John's crown in the shallow waters of the Wash,' explained the official Royal Navy report. 'On account of it being a very valuable prize,' the Americans dubbed it 'The Million Dollar Baby'.

It would be an extremely challenging operation. The site of the MiG wreck could only be reached through a shallow, narrow channel, mined at its southern entrance and within range of North Korean coastal batteries.

Spearheading the naval task force was the Royal Navy frigate HMS *Cardigan Bay*, supported by the light cruiser HMS *Kenya*, the carriers HMS *Glory* and USS *Sicily*, two US Navy landing craft and a South Korean motor launch. In addition to the naval personnel, several members of ATIC would take part in the salvage of the MiG.

On 20 July the first phase of the operation commenced, with a US Navy Dragonfly helicopter, operating from *Glory,* dropping a marker buoy at the crash site. The next day, when the tide was at its lowest, *Cardigan Bay* led the motor launch and landing craft through the 40-mile-long channel, anchoring as close as possible to the position of the wreck. With *Glory*'s Sea Furies providing low-level air cover and F-86s of the USAF's Fifth Air Force providing top cover, and the *Kenya* anchored in the Gulf of Yalu as a radar picket to give early warning of any approaching North Korean MiGs, the recovery operation got underway, as detailed in the RN report:

Ratings from the *Cardigan Bay* under Lieutenant M. Ross, RN set out in the ship's boats for the mudflats where they were joined by Americans from the landing craft, which had been fitted with a special crane.

Working half naked and without shoes, the party, some fifty-strong, found parts of the aircraft exposed on the mudflats. They lashed them with lines to which buoys were attached, so that the landing craft could lift them when the water rose over the flats. This was the worst part of the job, for it entailed floundering about among pools and depressions. The men must have been visible from the shore but said Captain Brown [captain of the *Cardigan Bay*]: 'They were very brave and worked as calmly and methodically as if they had been playing with their children on Margate Sands.'

With darkness falling and the tide going out again, the salvage team had to suspend the operation for the night, but they returned at first light the following morning to continue their work. However, the activity around the crash-site attracted the attention of the North Korean shore batteries, prompting intervention by the Corsairs of USS *Sicily*, as the report revealed:

> *Glory*'s aircraft had provided the air cover during the previous day but now aircraft from *Sicily* had taken over. They dived towards the shore and drew both light and heavy AA fire from batteries little more than a mile from the salvage operations. This caused some surprise.

Having recovered the fuselage and all other salvageable parts of the MiG, the naval task force withdrew, the *Cardigan Bay* firing a parting salvo from her guns at a North Korean battery as she did so.

> After the operation was over Captain Brown said that the part played by the aircraft from HMS *Glory* in locating and marking the position of the crashed fighter contributed materially to the success of the recovery and that the presence of the low flying close air support of *Glory* and *Sicily*'s aviators was of the greatest moral value. The work of the small boats with their mixed team of British, US and South Korean personnel was invaluable.[9]

Lieutenant General Otto Weyland, commander of the USAF's Far East Air Forces, also praised the personnel involved, sending a telegram extending his 'admiration and appreciation for [the] aggressive and determined spirit with which [this] difficult and dangerous operation was executed.'[10]

The Americans wanted to maximize the propaganda value of this successful operation by publicizing it in the press. But the British objected to this proposal, arguing that maintaining the operation's secrecy should take precedence. 'One of the ships taking part was fired on but it does not necessarily follow that the Russians are aware of the operation and its success. The North Koreans might be inclined to conceal their failure to prevent it,' Captain Patrick Brock, Director of the Royal Navy's Operations Division, pointed out. The office of the UK's Director of Naval Intelligence concurred: 'The advantages to be gained by this publicity would not outweigh the obvious gain to the enemy in knowing that we were in possession of one of his latest aircraft. He may well not be aware of it.'[11]

In the end, the Americans got their way and details of the operation were duly released to the media.

After being taken to Chodo Island, the 'Million Dollar Baby' was flown to Japan in a C-119 cargo plane and then shipped to Wright-Patterson AFB for extensive study by technicians of ATIC. The MiG wreckage, along with the components seized earlier by Donald Nichols' team, 'gave USAF intelligence the first real insight into Soviet aircraft, engine and precision production methods and the level of industrial technology attained by the Soviets' and 'added appreciably to the store of intelligence information elicited from other sources,' revealed an ATIC report.

The in-depth analysis of the MiG allowed the American experts to dispel certain myths that previously surrounded Soviet aircraft, including that the standard of manufacturing and quality of materials were inferior to Western aircraft. 'The MiG-15 denied some of our concept of Soviet workmanship heretofore thought to be generally inferior by US standards,' they admitted. 'Rubber and plastic parts from this aircraft which were examined proved that the Soviets were capable of good quality workmanship when it was really necessary in critical applications.' On the other hand, they found that 'poor work and materials were used in less critical parts', confirming the American view that the Soviets regarded their warplanes as 'a semi-expendable item.'

An area of particular interest to the Americans was the metals used in the MiG-15, which furnished ATIC's Metallurgy Group with 'new knowledge on the status of technology of aircraft metallurgy in the USSR.' The ATIC analysts were impressed by the quality of these metals and techniques of manufacture. 'Laboratory examinations of all the light metals used in critical areas reveal sound metallurgical and fabrication practices equal to and in some places superior to those in the US.' An example of this was the Nimonic alloy used in the VK-1 jet engine, which tests revealed to be 'slightly superior to its UK and US counterparts.' The VK-1, it was found, also used significantly less chromium and nickel than its Western equivalents. This, suggested the analysts, 'pointed the way to possible conservation of critical materials in the US.' Detailed examination of the VK-1 led the analysts to conclude that 'there was no appreciable lag in Soviet jet engine technology as compared to the US'[12]

Useful as the recovered wreckage was, what the USAF really needed was an intact, flyable example of the MiG-15 to undergo intensive flight testing. The job of securing this prize was given to the CIA.

The SOVMAT Program

To satisfy the growing demand from the military and various government agencies for items of Soviet hardware as Cold War tensions escalated, in July 1950 the CIA formed the SOVMAT Program, which had a remit to coordinate the collection of Soviet bloc materiel, both civilian and military, for technical analysis.

SOVMAT's field collection teams initially enjoyed little success, however. An official report put this down to a lack of manpower, rivalry with the military's own S & T collection units, and 'the dearth of opportunity to acquire or even to examine Soviet products' due to the fact that 'Soviet military equipment representative of Russia's best technological achievements is heavily guarded and rarely committed in combat where it is exposed to capture.'[13]

The call issued by the USAF for a flyable MiG-15 presented the CIA's stumbling SOVMAT Program with the perfect opportunity to prove its worth.

The Agency's first plan for meeting the Air Force's request was daring in the extreme. They would parachute a pilot into Poland, behind the Iron Curtain, who would then infiltrate a fighter base, steal a MiG-15 and fly it back to NATO territory. The man they selected for this highly ambitious undertaking was Polish exile Jozef Jeka.

A highly decorated Second World War veteran, Jeka had joined the Polish Air Force in 1937. Escaping to England after the fall of Poland two years later, he served with the RAF, flying Hurricanes during the Battle of Britain. By war's end he was a Squadron Leader, commanding No 306 Squadron, with many medals to his credit, including the DFM, the citation for which stated that he had shown 'the greatest courage and determination to inflict losses on the enemy.'[14] Remaining in the post-war RAF until 1949, he then became a contract pilot for the newly formed CIA, flying covert 'black' ops. Not only was Jeka an excellent pilot, he had the added advantage of being stateless, so the political fallout for the Americans would be limited should he fall into communist hands.

Jeka was sent to a secret location in West Germany to undergo training for the mission, where he would've learned everything about the MiG-15 ATIC had gleaned from the crashed example recovered from North Korean waters. However, in March 1953, shortly before the mission was due to be launched, an unexpected act by one of his fellow countrymen led to the operation being scrubbed.[15]

MiGs in Denmark

On 5 March 1953, Josef Stalin died. That day also marked another significant event in the history of the Cold War – albeit one inevitably overshadowed by the death of the Soviet dictator. On that morning, a MiG-15bis bearing Polish markings unexpectedly landed at Rønne airport, on the Danish island of Bornholm. Its pilot was 22-year-old Lieutenant Franciszek Jarecki of the Polish Air Force, who had flown from the Słupsk-Redzikowo air base, around 100 miles away, and requested political asylum upon landing. When news of the MiG landing reached the US embassy in Copenhagen, the air attaché immediately put in a request to the Danish government for permission to inspect Jarecki's MiG.

Denmark now found itself caught between the two superpowers – fearful of antagonising the Soviet Union, which was demanding the immediate return of both the aircraft and the pilot, but anxious not to alienate the US. Initially, the Danish defence minister Harald Petersen sought a compromise: Danish experts would examine the aircraft and share their findings with Washington. This, however, failed to placate the Americans, who pressed for their own technicians to be granted access to the aircraft. 'The [US] Department of State has instructed the American Embassy at Copenhagen to state that the United States Government considers the matter of "crucial military importance", and to seek a minimum assurance that the plane will not be returned promptly to Poland. Washington has suggested plans for examination of the plane by Americans technicians,' a US report stated.[16]

The British also requested that their experts be given access to the MiG. But with the Polish authorities adding to the pressure on Copenhagen to return the aircraft by seizing Danish fishing boats in the Baltic, Churchill's Cabinet observed that it might prove 'difficult to persuade the Danes to delay any further the return of the aircraft.' Nevertheless, they decided that 'every effort should be made to induce them to agree that it should first be inspected by British and American experts.'[17]

Eventually, the Danes relented. Inspections by US and British personnel would be permitted. But the Danes placed tight restrictions on these inspections, including that they would have to be completed by 16 March, the date Copenhagen had set for returning the MiG to

Poland. Furthermore, test flights would not be allowed. '[Copenhagen] indicated that this was the government's final position and stated that a "Danish facade" for examination of the plane must be maintained,' revealed a CIA report of 11 March, though it added that the American technicians were confident that 'a reasonably complete primary examination can be made within four days.'[18]

First to gain access to the aircraft was Wing Commander Francis Jeffs AFC, the air attaché at the British embassy in Copenhagen, who took many photographs of the MiG. He was followed by three Americans of ATIC. Having been thoroughly inspected, the MiG-15 was shipped to the Polish port of Gydnia. Jarecki, meanwhile, was taken to London, where he was celebrated as a hero by the anti-communist Polish Government in Exile, who awarded him the Polish Cross of Merit. He later moved to the US, providing the Americans with information on Soviet aircraft and the air tactics of the Warsaw Pact. As a reward for presenting NATO with a MiG-15, he was given $50,000 and granted US citizenship. Back in Poland, he was sentenced to death in absentia for treason by the communist authorities.

Jarecki wasn't the last MiG defector to land on Bornholm. Weeks later, on 20 May, another of his countrymen, Second Lieutenant Zdzislaw 'Jerzy' Jazwinski, made a wheels-up landing in his MiG-15bis at Rønne. The aircraft was damaged but Jazwinski was unhurt. According to a US intelligence report, Jazwinski had 'heard of Jarecki's defection over BBC and decided to defect in like manner.'[19] In July, he arrived in the US and like Jarecki was duly granted citizenship. Again, British and American technicians were granted access to the aircraft for inspection.

Thanks to the salvage operations in Korea and the Polish defections, the West had gained considerable insight into the MiG-15. But the ultimate prize – an intact, flyable example – still eluded the Americans. Efforts to finally remedy that situation focused once again on the Korean peninsula.

Operation Moolah

The new plan, devised by Far East Command's Psychological Warfare Section and codenamed Operation Moolah, was to entice a communist pilot to defect to South Korea by offering political asylum and a reward of $100,000 (around $1 million in today's money) to the first man to present them with an undamaged MiG-15. From April 1953, the offer was broadcast on radio in Korean, Chinese, Mandarin and Russian, promising 'refuge, protection, humane care and attention', as well as the huge cash reward. When the North Koreans began jamming the radio transmissions, USAF B-29s dropped over a million leaflets, designed and printed by the 1st Radio Broadcasting & Leaflet Group based in Tokyo, with the transcript of the broadcast over North Korea.

The armistice that brought an end to the fighting in Korea was signed on 27 July 1953 without any North Korean airman having taken up the American offer. Shortly after 10 am on 21 September 1953, however, a MiG-15 approached Kimpo AFB, just south of the 38th Parallel, the demarcation line between the two Koreas. The personnel at Kimpo were caught off-guard by the MiG's unannounced appearance, due to the base's radar being offline at the time for routine maintenance. The MiG landed on the airfield, narrowly missing a USAF F-86 Sabre which was coming in to land at the same time, and was immediately surrounded by military police.

Out of the cockpit stepped 21-year-old Lieutenant No Kum-sok of the 2nd Regiment, North Korean People's Air Force, who had taken off from his base at Sunan, near Pyongyang, 17 minutes earlier. Lieutenant Kum-sok was quickly hustled away for an initial round of questioning, before being transferred by helicopter to the headquarters of the 5th US Air Force Intelligence in Seoul, 12 miles away.

His aircraft, meanwhile, was sent to Kadena AFB in Okinawa. Here, US pilots finally got the opportunity to fly the fighter that had caused them and their allies so much trouble during the Korean War. Among those who flew it was the famous test pilot and first man to break the sound barrier Major Chuck Yeagher. It was then dismantled and transported to Wright-Patterson AFB for a more intensive programme of test flights.

The ATIC experts confirmed that the Soviet machine had a higher ceiling and faster rate of climb than the F-86 but possessed inferior handling. They also praised the cockpit arrangement, finding that, when strapped into his seat, the pilot 'can reach all the controls, instruments, indicators, and emergency equipment with ease', although the rear view through the cockpit canopy was found to be poor.[20]

While the MiG-15 was being put through its paces, No Kum-sok spent the next six months being debriefed by intelligence and USAF officers. Besides giving his interrogators a detailed insight into the inner workings of the North Korean Air Force and communist air combat tactics, he revealed the full extent of Soviet involvement in the Korean air war. He also told his American hosts that his defection had been motivated by his growing disenchantment with the North Korean political system, rather than the reward money, of which he claimed to be unaware at the time of his defection flight. He received the $100,000 and US citizenship, later changed his name to Kenneth Rowe and enjoyed a successful career in the US aerospace industry before retiring to Florida.

Chapter Six

Unexpected Guests

Franciszek Jarecki's unexpected arrival on Bornholm in his MiG-15, and the diplomatic tug-of-war that ensued over the return of the aircraft, prompted the RAF to seek guidance from the Air Ministry 'on the procedure to be adopted, if a Soviet or Satellite aircraft should land unexpectedly on a British airfield', whether any such landings should be intentional or accidental.

J.S. Orme of RAF Air Intelligence wrote to the Air Ministry on 30 April 1953:

Presumably, if a Russian or a Chinese or a satellite aircraft landed on an RAF airfield in peacetime, our aim would be to keep control over the aircraft and its crew as long as we could. The main decisions about how this should be done and for how long it could be done, should, if possible, be taken in London (though the problem might be rather different in respect of aircraft landing on airfields which were under the operational control of SACEUR).

I am inclined to think that we could almost accept that, as a matter of commonsense, any RAF units confronted with this problem would refer to higher authority and that the incident would very quickly be reported to London; meanwhile, the unit could probably be relied on to take administrative measures to hold the aircraft and aircrew. It might, however, be worthwhile issuing instructions to this effect to RAF Commands other than those under SACEUR's operational control, and inviting SACEUR to take similar action.[1]

Another RAF officer argued that, in the event of a Soviet aircraft landing by accident at an RAF base in West Germany, the personnel at the airfield should take action 'to prevent the pilot taking off in a hurry when he had realized his error.'

On 7 May 1953, an Air Ministry official added his own thoughts, suggesting that, if a defecting crew landed and Moscow demanded the aircraft's immediate return, they could 'indulge in endless delaying tactics (e.g. taking time to issue passes to the people sent to collect the aeroplane).' The Soviets, he added, 'seem to have few inhibitions on such occasions.' However, the man from the Ministry cautioned that they should 'probably think twice about trying to hang on to the aircraft permanently – even if you had an English lawyer's opinion that International Law countenanced such action, which is more than doubtful.'

The question of possible Soviet landings at civil airports in the UK was also examined, though it was felt that it would be 'unnecessary to issue special instructions regarding Soviet and Satellite aircraft landing at civil airfields, since it seems likely that the police and civil aerodrome authorities already have general procedures for dealing with unexpected or irregular landings.'

Ultimately, the Air Ministry's guidance was that, in the event of an unexpected landing on an RAF base by a Soviet bloc aircraft, the senior officers at the base should 'detain the aircraft as long as possible, and obtain instructions from higher authority about its disposal. SHAPE would treat each case on its merits, and deal direct with the country concerned.'[2]

Even before Jarecki's flight to Denmark, there had been surprise landings by Communist airmen in Western Europe. On 9 October

1948, the crew of a Tupolev Tu-2 light bomber, pilot First Lieutenant Anatoly Barsov and his navigator, Second Lieutenant Pyotr Piragov, flew from Kolomyia in western Ukraine to the US air base in Hörsching, Austria. Barsov and Piragov were granted asylum in the US, while the third member of the crew, who did not wish to defect, was repatriated. The Tu-2 was later returned to the Soviets. Another Second World War-era Soviet warplane, an Ilyushin Il-10 *Sturmovik*, was delivered into American hands on 28 April 1950 when North Korean pilot Lieutenant Lee Kun-Soon defected to the South.

On the other side of the Iron Curtain, the Soviets were benefitting from accidental landings by NATO airmen on the territory in their zone of occupation. One of the first such incidents occurred on 16 August 1948, when the pilot of an RAF de Havilland Vampire F.1 of No 3 Squadron, based at Lubeck, became lost during a training sortie. Running low on fuel, he made a forced landing in East Germany. 'Later in the day news was received through Intelligence sources that the aircraft had "belly" landed and the pilot was seen standing by his machine surrounded by Russians. The pilot appeared to be unhurt,' the Squadron ORB recorded.[3]

The next day the ORB reported that news had reached the squadron that the pilot 'was being well looked after by the Russian authorities and that we could go into the [Soviet] Sector to inspect the aircraft and pilot.' The pilot arrived back at his base, unharmed, five days later.

Not all such incidents were resolved so speedily, however. A major behind-the-scenes diplomatic battle between London and Moscow ensued in 1950, when on 5 September a Gloster Meteor T.7 trainer jet, on a ferry flight from RAF Eindhoven to Wunstorf, made an emergency landing in a field near the town of Redefin, 32km across the East German border, after the pilot, Flight Lieutenant John Driver, became lost in bad weather. Flight Lieutenant Driver was taken into custody by the Soviets, who refused to allow any British officials to visit him or inspect the aircraft. It soon became clear that Moscow was linking the release of the pilot and his Meteor to the case of a junior Soviet Army officer who had recently defected to the British zone of West Germany, whom they were demanding be handed back.

A memo of 15 September 1950 from an aide to the Secretary of State for Air Arthur Henderson states:

> The Foreign Office and the Air Staff are very averse to the return of the Russian deserter (an Army Lieutenant who has recently been in Russia) because: a) we expect to get a good deal of valuable intelligence from him; b) If he is returned, he will undoubtedly be executed. This will become known, and will effectively prevent any further Russian desertions – which we are very anxious to encourage for intelligence reasons . . . The UK High Commission are trying to keep the two cases quite separate, and are seeking the return of [Flight Lieutenant Driver] without themselves returning the Russian deserter.[4]

However, Henderson's aide conceded that, should negotiations prove fruitless, 'in the last resort we may have to consider accepting the Russian bargain.' The return of the Meteor was, he added, 'relatively a secondary matter,' though 'obviously we should get it back if we can.'

Secret negotiations dragged on for weeks. On 7 October, the Soviets made a minor concession by permitting an RAF Wing Commander serving with BRIXMIS – the British military liaison mission that operated in the Soviet zone of Germany – to inspect the crashed Meteor, which hadn't been moved from the crash-site, under close Soviet supervision. The Wing Commander reported on the damage:

> Port oleo torn off. Starboard wing tip partly buried, port wing torn off at root, small hole in leading edge of starboard mainplane but impossible to ascertain what caused it. Canopy had been jettisoned by Pilot, probably after aircraft came to rest. Engines

apparently undamaged. Aircraft does not appear to have been tampered with and is obviously in its original position.

'Damage to foreign property,' he added, 'consists of 1 cow killed and slight damage to root crops.'[5]

The diplomatic impasse continued for several more weeks. By the end of October, with no sign of the Soviets willing to budge, the pilot's case was taken up by Prime Minister Clement Attlee and Foreign Secretary Ernest Bevin. This top-level intervention eventually bore fruit. On 24 November 1950, after 80 days in Soviet captivity, Flight Lieutenant Driver was finally handed over to British officials at the Hamelin Bridge checkpoint in Lower Saxony. The Meteor T.7 wreck was retained by the Soviets.

Almost four years later another Gloster Meteor, this one a two-seat, radar-equipped night fighter variant, the NF.II, belonging to the RAF's No 527 Squadron, made an emergency landing in the Soviet zone of East Germany after running out of fuel. On this occasion the matter was resolved swiftly. 'The crew were returned to the British authorities after approximately 4 days in Russian hands and appeared to have had quite a good time,' recorded the Squadron ORB.[6] The Meteor NF.II (serial WM 1884), which had suffered Category 3 damage in the crash, was also returned.

The 1960s saw several incidents of Warsaw Pact airmen putting their aircraft down on West European territory, both by accident and on purpose. On 4 August 1961, the West finally got its first look up close at the MiG-17 when Senior Lieutenant Ludwig Slamal of the Czechoslovakian Air Force made an emergency landing at an airport in the Austrian capital, Vienna. Introduced into Soviet service in 1952, the MiG-17 was an improved version of the MiG-15. Among the improvements it incorporated were thinner wings swept back at a greater angle, helping raise the maximum speed to 680 mph, and more docile handling characteristics. After a brief inspection, the MiG-17 and its pilot were returned to Czechoslovakia.

A second MiG-17 fell into NATO's hands six years later, this time as a result of a deliberate defection. On 25 May 1967, VVS pilot Lieutenant Vasily Epatko took off from his base at Dubno in western Ukraine. Lieutenant Epatko soon separated from the rest of his flight and flew across East Germany. Penetrating West German airspace at low altitude, he made a wheels-up landing in a meadow in the Bavarian town of Kicklingen. The pilot then hitched a lift to the nearest military base, where he requested political asylum. No fighters were scrambled to intercept the intruder, and the fact that Epatko was able to successfully breach West German airspace to a depth of 100 miles without provoking a response from NATO's air defences was a source of considerable embarrassment to US commanders in Europe. 'All hell has broken loose, with three-star generals fighting like cats and dogs. Army intelligence is in turmoil,' a US military source disclosed to the press at the time.[7]

Despite the discomfort suffered by the NATO top brass by his unexpected arrival, Lieutenant Epatko was granted asylum in the US. Taken to a CIA safe house on the shores of the Chesapeake River, he was interviewed by, among others, defence analyst and aircraft design consultant Pierre Sprey.

Outside of Europe, yet another MiG-17 was presented to the Americans courtesy of Cuban Air Force defector Lieutenant Eduardo Jiminez, who flew his fighter to Homestead Air Force Base in Florida on 5 October 1969. The MiG-17 was later returned to Cuba.

A much greater prize slipped through NATO's fingers on 13 February 1967. During a ferry flight from Minsk to East Germany in a MiG-21PFM, the latest variant of the interceptor, Captain Zinovyev of the 16th Guards Fighter Aviation Division inadvertently crossed into West Berlin. Mistaking Tegel Airport in the French sector for his intended base in Cottbus, he put the fighter down. Seeing a row of parked aircraft bearing *Armée de l'Air* markings, Zinovyev realized his error and hastily took off again before personnel at the airport could block the runway with vehicles.

Of course, Soviet strategic bombers were of special interest to Western intelligence. A rare opportunity to gain first-hand information on an aircraft that, along with the Tu-95 'Bear', formed the backbone of the Soviet DA fleet came NATO's way in 1968. On 25 May the aircraft carrier USS *Essex* was conducting ASW exercises in the North Sea, off the Norwegian coast, when a Tu-16 'Badger-F' of the AVMF, configured for ELINT duties and operating from the Severomorsk-3 air base in Murmansk, made several low passes near the *Essex*, at an altitude of 120ft. On its fourth pass, the pilot Colonel Andrei Pliyev lost control and crashed into the water. None of the six crew survived.

An SH-3 rescue helicopter from the *Essex* recovered the bodies from the water, which were handed over to the crew of a Soviet destroyer that was in the vicinity – though not before the Americans had retrieved 15 documents the crew had been carrying. These were translated and, after being examined by the Americans, passed on to the British for analysis by the DSTI, which was responsible for assessing S & T intelligence on Warsaw Pact equipment.

'Evidence in these documents, and the mark of BADGER involved, suggest that the crew was briefed, and the aircraft equipped, for Radar, Radio and Photographic Reconnaissance,' the analysts confirmed. Although most of the documents were routine papers, such as check lists, a flight plan and the navigator's log sheets, which, the analysts noted, 'would be expected as the stock in trade of a maritime reconnaissance air crew,' they also included details of the tactics Soviet bomber crews were to employ against NATO shipping in time of war, and so had considerable intelligence value.[8]

One of the more dramatic defection flights from the Soviet bloc occurred on 27 May 1973, when Lieutenant Yevgeny Vronsky, a

20-year-old aircraft technician serving with the 296th Fighter-Bomber Aviation Regiment, defected to the West in a Sukhoi Su-7 he stole from his base in Grossenhain in Saxony. Lieutenant Vronsky's flying experience was limited to a few sessions in a simulator, and he made the entire flight in afterburner mode and with the undercarriage lowered. After running out of fuel, he ejected near the town of Braunschweig, the Su-7 crashing into a forest. West German Chancellor Willy Brandt granted Vronsky political asylum and the Su-7 wreckage was returned to the Soviets.

Viktor's flight to freedom

Undoubtedly the most famous and, from an intelligence standpoint, important pilot defection of the Cold War took place on 6 September 1976. Shortly before one pm, a flight of five MiG-25Ps (the P standing for *Perekhvatchik* – 'interceptor') took off from the Chuguyevka air base in Primorsky Krai, 250 miles northeast of Vladivostok, on a training sortie.

A large, powerful interceptor, the MiG-25 was designed in the early 1960s to defend Soviet airspace from the North American XB 70 'Valkyrie', then under development in the US and expected to form the backbone of SAC's nuclear-armed bomber fleet from the 1970s. Intended to be capable of flying at Mach 3 at an altitude of 70,000ft, the XB-70 would've presented Soviet air defences with a formidable challenge. Cost overruns and a re-evaluation of tactics by SAC commanders following the shooting down of Francis Gary Powers' U-2 in May 1960, who opted for a switch from high to low-level penetration of Soviet airspace, led to the cancellation of the XB-70 in 1961. But the inflexible nature of Soviet defence procurement meant that development of the MiG-25 continued, and the prototype made its maiden flight on 6 March 1964. Built partly of titanium and powered by two Tumansky R-15B-300 turbojets, it could reach Mach

2.83 in level flight and had a service ceiling of 89,000ft. The MiG-25 would set 29 aviation records, most for speed and altitude, several of which stand to this day. Besides its blistering performance, the MiG-25 was also heavily armed, carrying four radar and infra-red guided air-to-air missiles, and fitted with a powerful Smerch-A2 fire control radar in its nose capable of detecting targets up to a range of 100km.

Although the XB-70 had been cancelled, new impetus was injected into the MiG-25 project when the USAF introduced the Lockheed SR-71 'Blackbird' strategic reconnaissance aircraft in 1966. With a maximum speed of almost 2,200mph and capable of reaching an altitude of 85,000ft, only the MiG-25 had any chance of intercepting this formidable new spy plane. 'The MiG-25 program was also kept going for technology derived from it,' according to a CIA report written towards the end of the Cold War. 'Perfected welding techniques and steel fabrication processes under the program are still being utilized throughout heavy industry in the USSR.'[9]

Unsurprisingly, when it entered service with the PVO in 1970, the MiG-25 was regarded as a major threat by NATO, which assigned it the reporting name 'Foxbat'. Its capabilities were demonstrated in 1971, when reconnaissance variants successfully evaded Israeli F-4 Phantoms while conducting PR missions over the Israeli-occupied Sinai Peninsula on behalf of the USSR's then ally, Egypt.

Among the pilots who took off from Chuguyevka on 6 September 1976 was 29-year-old Lieutenant Viktor Belenko. Born in the city of Nalchik in the Caucasus in 1947, Belenko graduated from the Soviet Air Force's Higher Military Aviation School in Armavir, Krasnodar Krai. In 1974 he was assigned to the 513th Fighter Regiment of the PVO, which was equipped with the MiG-25P, at Chuguyevka air base, where their primary task was to intercept any Lockheed SR-71 'Blackbird' reconnaissance aircraft conducting spy flights at extreme altitude over the Soviet Far East.

Lieutenant Belenko, however, had become increasingly disillusioned with the repressive Soviet regime, later describing it as a 'corrupt, abusive system', and made up his mind to defect in his Foxbat to Japan, the nearest non-communist state to his base. Minutes after taking off from Chuguyevka, whilst at an altitude of 19,000ft, Belenko put his MiG-25 into a steep dive. The other Soviet pilots watched with alarm as the big MiG disappeared into cloud. Unable to raise him on radio, the remaining four Foxbats sadly headed back to base, assuming that their unfortunate squadron mate must have suffered a mechanical malfunction and crashed into the sea.

Belenko, in fact, had pulled out of his dive at an altitude of just 100ft. Switching off his radio and radar, he set course for Chitose air base on Hokkaido, the northernmost of the Japanese home islands, maintaining an altitude of just 200ft to avoid being picked up by Soviet radar. Once out of Soviet radar range, Belenko gained height to conserve fuel and, when around 200 miles off the coast of Hokkaido, was spotted on Japanese radar. After attempts to raise the unidentified aircraft on radio failed, two F-4 Phantoms of the Japanese Air Self Defence Force were scrambled to intercept but were unable to track the incoming intruder.

As Belenko approached the Japanese coast, a low-fuel warning light started flashing in his cockpit. Realizing he wouldn't make it to Chitose, he diverted to the nearer civil airport at Hakodate. Forced to abandon his first attempted landing due to a Boeing 737 taking off, Belenko circled around and successfully put the MiG down on his second attempt. However, as he thundered down the tarmac, the pilot realized the runway was too short for his big interceptor and, despite deploying the aircraft's braking chutes, he overshot the runway by some 200ft before the MiG finally juddered to a halt.

After overcoming their initial shock at the unannounced arrival of a Soviet fighter, several airport personnel cautiously approached the MiG. Belenko opened the canopy and warned them off by firing two rounds in the air from his service pistol. The tense stand-off that ensued finally ended when an official at the airport warily approached waving a white flag and was handed a note by Belenko, which read: 'Quickly call representative American intelligence service'. Realizing that they were dealing with a defection rather than simply a lost pilot, the airport's managers summoned an interpreter. After requesting political asylum in the US, Belenko was flown by police helicopter to his original intended destination, Chitose air base, before being moved on to Iruma air base just outside Tokyo, for questioning. He was then charged with unauthorized entry into the country and illegal possession of a firearm.

News soon reached the US embassy that a Soviet pilot had landed in Japan in a MiG-25 claiming asylum, and the Americans lodged a request with Tokyo to interview the pilot. At the same time, Moscow demanded that Lieutenant Belenko, whom they insisted to the world's press was being held against his will, be handed over to their officials at the Soviet embassy.

Wary of angering the Soviets, the Japanese permitted one of their embassy officials to talk to Belenko. The meeting, which lasted seven minutes, did not go well. The official – who was also the embassy's resident KGB officer – allegedly called the pilot a traitor when he rejected an offer of repatriation to the USSR. CIA officers who arrived in Japan were also granted access to Belenko. Three days after his arrival at Hakodate airport, the Soviet pilot was flown to a USAF base close to the CIA's HQ in Langley, Virginia.

Tokyo's refusal to hand back Belenko and his MiG provoked a furious response from Moscow. 'The acts of [the] Japanese authorities with regard to the Soviet plane and its pilot could not be qualified as being other than unfriendly to the Soviet Union, flouting elementary norms of international law and the practices of relations between states, especially neighbour states,' fumed a Soviet government spokesman to the TASS news agency.[10]

Meanwhile, on 13 September, President Gerald Ford publicly confirmed that Belenko's request for political asylum in the US would be approved. 'As long as he wants such asylum,' he announced during a press conference, 'he will be granted it in the United States.'[11] In response, the Soviets held a televised press conference, at which Belenko's wife and mother both tearfully pleaded for his return, assuring him that the authorities would forgive his 'mistake' in defecting. This appeal had no effect, however, and Belenko remained in America, where he was exhaustively interrogated by a team of CIA officers.

'A major intelligence bonanza'

The director of the CIA – and future US President – George H.W. Bush described Belenko's defection as 'a major intelligence bonanza.'[12] The pilot provided his hosts with a wealth of intelligence on the inner workings of the PVO, and revealed that conditions at the Chuguyevka air base were poor and the morale of the personnel stationed there low. Of particular interest to the CIA was his disclosure that a more advanced version of the MiG-25 was currently under development, designated the MiG-31. Initially referred to in the West as the 'Super Foxbat' before receiving the official reporting name 'Foxhound' by NATO, the MiG-31 was a long-range interceptor with a two-man crew. Belenko told the CIA that the MiG-31 was to have uprated engines, improved avionics and be armed with a more advanced, longer-range AAM. The defector's most disturbing revelation was that the MiG-31 would be equipped with a new radar that would have full lookdown/shootdown capability, a technology which the USSR was thought to trail far behind the US, and which would make the MiG-31 a potent threat to low-flying NATO aircraft. The MiG-31 'Foxhound' eventually entered PVO service in 1982.

While Belenko was being debriefed by the CIA, on 25 September his MiG-25 was transported aboard a USAF C-5 Galaxy from Hakodate to Hyakuri air base. Resisting Soviet diplomatic pressure for the immediate return of the aircraft, the Japanese agreed to American requests for their experts to be granted access to the MiG. A team from the Foreign Technology Division at Wright-Patterson AFB arrived at Hyakuri, where they dismantled the Foxbat and carried out a detailed inspection.

The American experts soon discovered that the MiG-25's capabilities had been somewhat overestimated in the West. Although certainly fast, its avionics and Smerch-A2 radar, which utilized vacuum tube technology, were relatively crude by Western standards. It also became clear to the analysts that, rather than a pure fighter, the 'Foxbat' was essentially an interceptor, with poor manoeuvrability.

The MiG-25's technical inferiority was remarked upon by US Secretary of State Henry Kissinger when he met with the Chinese Foreign Minister Chiao Kuan at the UN in October 1976. 'As for [the Soviets'] strength, the latest plane that we got in Japan shows that they are really quite backward. The plane is about 10 per cent better than our planes of 14 years ago. If this achievement is the result of a high priority project in the Soviet Union,' Kissinger mocked, 'I hate to think of the outcome of their low priority projects.'[13]

Despite the Americans' generally poor opinion of the Foxbat, a secret CIA report from 1990 revealed that by the 1980s the MiG-25 posed a serious threat to SR-71 missions being conducted close to and over the Soviet Union. 'SR-71 incursions into restricted airspace over the Soviet Union were often rebuffed by the launch of MiG-25s which forced SR-71s to retreat from their surveillance patterns over Soviet territory,' the report disclosed.[14]

At a heated meeting with President Ford and Kissinger in the Oval Office on 1 October 1976, the Soviet Foreign Minister Andrei Gromyko made clear his government's displeasure. '[The] US and

Japan took the plane apart as if they owned it, like spoils of war. We can't qualify this as anything but hostile,' he fumed. 'General Secretary Brezhnev told us to tell you he is very bitter and indignant', adding that the affair 'has caused great indignation in the whole country.'[15]

After the US experts had completed their inspection, the dismantled parts of Belenko's Foxbat were crated and shipped back to Vladivostok aboard a freighter in November 1976. According to a CIA report, the Soviets seriously considered cancelling further production of the aircraft, as its secrets were now known to NATO. As it was the only fighter in PVO service capable of intercepting the SR-71, however, the Foxbat earned a reprieve.

Viktor Belenko was sentenced to death in absentia for treason by the Soviets. Fearing the KGB might try and assassinate him, he was given a new identity by the CIA and his location was kept a closely guarded secret. In 1980 he became a US citizen and married an American woman, with whom he had two sons. He also prospered professionally, enjoying a successful career as an aerospace consultant and co-authoring an autobiography, *MiG Pilot*, which detailed his defection.[16]

There have been rumours that Belenko's defection flight may not have come as a complete surprise to the Americans. Days after he landed at Hakodate, West Germany's *Stern* magazine carried a report claiming that an Austrian engineer working for the CIA had been in contact with Belenko and had persuaded him to defect. Five years later another newspaper report further fuelled rumours of Western intelligence involvement, quoting an unnamed Japanese intelligence source who insisted that Belenko had been recruited by SIS almost two years before his flight and that his defection had been carefully planned by both British and American intelligence. Asked to comment on this report, CIA spokesman Dale Peterson would only say: 'We would not comment one way or the other on any allegations of operational activity.'[17] In the absence of any firm evidence, claims that

Belenko's defection in the MiG-25 was the result of a joint SIS-CIA operation remain unproven.

Project Have Boat

It wasn't just Soviet bloc pilots who were providing the West with an intelligence windfall by defecting in their aircraft at this time. On 7 July 1977, a squadron of Shenyang J-6 fighters of the PLAAF took off from their base at Jinjiang in Fujian Province for a routine patrol. Once over the Taiwan Strait, the squadron's commander, 41-year-old Fan Yuanyan, separated from the rest of his Flight and headed for Taiwan. Realizing he was trying to defect, his squadron mates chased him into Taiwanese airspace, apparently with the intention of shooting him down, and only broke off when Taiwanese fighters arrived to escort the runaway J-6 to Tainan air base in southern Taiwan. Once safely on the ground, Yuanyan explained to the Taiwanese officers at the base that the miserable conditions on mainland China had motivated his defection. While the pilot was taken to the capital Taipei for a thorough debriefing, a team of USAF and ROCAF personnel carried out a joint examination of the J-6 at Tainan, under the codename Project Have Boat.

The Shenyang J-6 was the Chinese-built copy of the MiG-19 'Farmer'. Distinguished by its highly swept wings, the MiG-19 entered VVS service in 1955 and the USSR's Warsaw Pact allies soon after. Heavily armed with three 30mm cannon, and powered by a pair of Tumansky RD-9B engines, it was notable for being the Soviet Union's first operational fighter capable of supersonic speed (and the world's first mass-produced supersonic fighter). The essentially identical J-6 was built by the Shenyang Corporation from 1958 until 1981, around 4,500 being produced. Widely exported, the J-6 saw action in the Six Day War, the Indo-Pakistani War of 1971, and with the VPAF in the latter stages of the Vietnam War. American pilots who encountered the 'Farmer' reported that it seemed more manoeuvrable than the more advanced MiG-21. While the MiG-19's service life with the VVS was relatively short (being withdrawn in the early 1960s), the J-6 enjoyed a far longer career with the PLAAF, the last examples only finally being retired in 2019. The J-6 also served as the basis for the locally-built Nanching Q5 'Fantan' strike aircraft, which entered PLAAF service in 1970.

Although more or less obsolete by 1977, as the most numerous frontline fighter in PLAAF service, the offer to examine Yuanyan's J-6 was enthusiastically taken up by US intelligence. After their initial joint examination with the Taiwanese, Lieutenant General Harold Aaron, Deputy Director of the DIA, reported to the Director of the CIA that they had 'gained extremely valuable first-time insight into the PRC scientific and technical capability to produce a complex weapons system.' The Taiwanese even offered to allow the Americans to remove the J-6's Liming Wopen-6A engines, IFF system and cannon, and secretly ship them back to the States for a more thorough analysis. 'Detailed examination and operation under laboratory conditions of these components would substantially enhance the value of information obtained during field exploitation. The majority of the components would be functionally tested, then disassembled and examined for manufacturing technologies and factory markings, the latter identifying manufacturing locations and rates of production,' wrote Lieutenant General Aaron.[18]

Twelve years later, on 11 October 1989, a Syrian Air Force MiG-23MLD penetrated Israeli airspace at very low altitude and landed at a civil airport near the town of Megiddo in northern Israel. While the Israeli government was left somewhat embarrassed by the ease with which a lone MiG from a hostile country had managed to breach Israel's famously formidable air defences, the unexpected arrival of a MiG-23 was warmly received by the Israeli Air Force.

Initially, the Syrian government claimed the pilot had been forced to land in Israel after experiencing mechanical problems during a training flight, and was being held against his will. But days later a press conference was held in Israel at which the pilot, Major Mohammed Bassem Adel, confirmed that he had intentionally defected to Israel, motivated by a desire to live in a democratic country.

The MiG-23 (known to NATO as 'Flogger') was developed in the 1960s as the successor to the MiG-21, aiming to rectify the earlier type's lack of endurance and modest armament. The first Soviet aircraft to feature variable-geometry 'swing' wings, the prototype of the MiG-23 made its maiden flight on 10 June 1967, with production beginning three years later. Entering service with the VVS in 1971, besides the main interceptor variant, a ground-attack version was also developed, which eventually matured into the MiG-27. Although an ageing design by 1989, the 'Flogger' still served in significant numbers in the air forces of several of Israel's traditional foes, and Major Adel's defection was therefore considered 'an important event for the [Israeli] Air Force,' by General Mordechai Hod, a former commander of the IAF, who added: 'The significance is that the Air Force can now examine the plane secretly and test its electronics.'[19]

The MiG-23 was taken to Tel Nof AFB, where IAF test pilots carried out more than a dozen evaluation flights. Among their findings were that the 'Flogger' had greater acceleration and a more capable radar than previously thought. Israeli technicians were also able to unlock the secrets of the MiG's ECM equipment. After completion of the test programme, Major Adel's MiG-23 was put on display at the IAF museum at Hatzerim.

Flight of the Fulcrum

The final Soviet pilot defection of the Cold War occurred just six months before the fall of the Berlin Wall, and was one of the most dramatic of all.

Like Viktor Belenko, Captain Aleksandr Zuyev, a 28-year-old pilot serving with the 176th Guards Fighter Aviation Regiment based at Gudauta in Georgia, had become deeply disillusioned with the Soviet system. He was also experiencing marital difficulties. In May 1989, after being passed over for a coveted position as a test pilot, he made up his mind to defect to neighbouring Turkey in a MiG-29.

Known to NATO as the 'Fulcrum', the MiG-29 was a lightweight, twin-engine supersonic tactical fighter, development of which began in the early 1970s to replace the ageing MiG-21, MiG-23 and Sukhoi Su-17. It began entering service with the VVS from 1982 and later equipped all the Warsaw Pact air forces. It was also widely exported to the Soviet Union's client states. The West got its first look up-close at the MiG-29 when two examples made an appearance at the Farnborough air show in England in September 1988, as part of Mikhail Gorbachev's policy of *glasnost*. However, the two MiGs were closely guarded by the Soviets during their stay in the UK, preventing any inspection of the aircraft.

Captain Zuyev put his defection plan into action on the evening of 20 May. After subduing most of the personnel at his base by giving them slices of cake to celebrate his wife's pregnancy laced with crushed sleeping pills, he cut the base's telephone lines and made his way to one of two flight-ready MiG-29s. As he tried to gain access to his chosen MiG, however, he was challenged by a sentry and a struggle ensued, during which shots were fired and both men injured, Zuyev being hit in the arm. Nevertheless, he still managed to take off in the MiG-29 and headed across the Black Sea to Turkey. A Soviet account states that another of the squadron's pilots took off in pursuit of Zuyev in the other MiG-29, but that he was unable to maintain the chase due to the effects of the spiked cake he'd eaten. Disregarding a refusal of permission to land by Turkish air traffic control, Zuyev put his MiG down safely at Trabzon air force base in northeastern Turkey.

After opening the cockpit, he reportedly declared 'Finally, I am an American!', before being taken to hospital, where he requested asylum in the US.[20]

The Turkish ambassador in Moscow was summoned to the Soviet Foreign Ministry, where an official demanded the immediate return of both the aircraft and Captain Zuyev. Within hours of Zuyev's landing, the Chairman of the US Joint Chiefs of Staff, Admiral William J. Crowe, contacted the Chief of the Turkish General Staff, General Necip Torumtay, to request access to the MiG-29, assuring him he had a team on stand-by in West Germany ready to fly to Trabzon at a moment's notice. Although a NATO member and close ally of the US, Turkey was anxious not to provoke the Soviets, especially as East-West relations were greatly improving under Gorbachev's more moderate leadership. Thus, requests from both Washington and London to inspect Zuyev's aircraft were rejected by Ankara.[21] 'The Turkish government wants to maintain good ties with the Soviet Union. Our governments have agreed that a team of Soviet airmen will go to Trabzon and bring the aircraft back to the Soviet Union,' a Turkish diplomat explained to the press.[22]

The MiG-29 was returned to the USSR on 21 May. As for Zuyev, he was charged with hijacking an aircraft, but the charges were later dropped as his actions were judged to be political rather than criminal. In June 1989 his request for political asylum in the US was granted and, like Viktor Belenko before him, he co-authored a book about his defection. He was killed in a flying accident in June 2001.[23]

Chapter Seven

The Coldest Years Part I
Soviet Bloc Intelligence Operations 1953 – 1969

After the death of Josef Stalin in March 1953, there were hopes that the end of the tyrant's rule would herald a new, more constructive relationship between the USSR and the West. These hopes were soon dashed, however, with the next decade and a half being the most tense and dangerous years of the entire Cold War. Under the leadership of Stalin's successors, the erratic Nikita Khrushchev and the hardliner Leonid Brezhnev, the period 1953 to 1969 would witness multiple popular revolts against communist rule within the Soviet bloc, from the East German Uprising of June 1953 to the Prague Spring of January – August 1968, major East-West confrontations over the erection of the Berlin Wall in 1961 and the Cuban Missile Crisis of 1962, as well as deepening US military involvement in Vietnam.

Against this backdrop of high tension, the nuclear threat escalated dramatically, as both sides built up vast nuclear stockpiles of ever greater destructive power. In November 1952, the Americans tested the world's first hydrogen bomb. Codenamed 'Ivy Mike', this device had a yield of 10.4 megatons, many times more powerful than the atomic bomb dropped on Hiroshima. The Soviets were quick to join the 'H-bomb' club, detonating their first example just nine months later.

Parallel to the development of new nuclear weapons was the introduction of a new generation of more potent, mostly jet-powered bombers to carry them. The USAF formed SAC (Strategic Air Command) in 1946. Initially equipped with wartime B-29s, SAC soon grew into a massively potent force, fielding over 2,500 bombers by the late-1950s, the most important of which was the formidable Boeing B-52 'Stratofortress'. SAC's Soviet equivalent was the DA, an independent arm of Soviet military aviation, which was also established in 1946. DA gained its first true strategic heavy bomber in 1949 with the Tu-4 'Bull'. But the Tu-4 was not an ideal nuclear bomber for the DA, lacking the range to strike the Continental United States from Soviet air bases. Furthermore, by the time it had entered service, the piston-engined 'Bull' was already becoming obsolescent, the type's vulnerability to modern jet interceptors being exposed by the Soviets themselves during the Korean War, when USAF B-29s were mauled by MiG-15s.

Long-range, jet-powered strategic bombers were clearly needed as carriers of the Soviet Union's nuclear weapons, and these were not long in arriving. In April 1954 the twin-engined Tupolev Tu-16 'Badger' became the first jet-powered strategic bomber to enter service with the DA. Next came the four-engined Myasishchev M4 'Bison', which entered service in 1955 (albeit in very small numbers). The most famous and longest-serving of this new generation of bombers was the Tupolev Tu-95 'Bear', which began equipping DA units from 1956. Unlike most bombers of that era, rather than jets the mighty 'Bear' was powered by four giant turboprop engines with contra-rotating propellers. Nevertheless, the power produced by the Kuznetsov NK-12 units, coupled with the swept-wing design, ensured the Tu-95's performance was comparable to its jet-powered contemporaries.

The appearance of these aircraft in flypasts at Soviet May Day air displays in the mid-1950s sparked considerable alarm in military circles in the West, no more so than in the US, who believed that the Soviets now possessed nuclear bombers with global reach – though, in fact, only the Tu-95 had the range to attack the Continental United States from Soviet territory on a round trip. The response of NATO and other Western air forces to the increased Soviet bomber threat was to develop new, supersonic interceptors to defend their nations' airspace. These projects became an important target for Soviet intelligence.

The 1950s, however, was a difficult period for Soviet intelligence in the US. Not only had the Venona codebreaking programme successfully unmasked many of their agents in America, but the 'Red Scare' propagated by Senator Joseph McCarthy had fostered a culture of intense suspicion and vigilance against communist infiltration of American society. FBI counterintelligence had also become more adept at identifying and 'turning' the NKVD's American assets.

One such 'turned' NKVD spy was Boris Morros, one of the more colourful characters recruited by Soviet intelligence. Born in St Petersburg in 1891, Morros emigrated to the United States in 1922 and went to work in Hollywood, joining Paramount Pictures. He eventually became a modestly successful film producer, one of his biggest hits being Laurel and Hardy's *The Flying Deuces*. On a trip back to Russia in 1934 he was recruited by the NKVD. Given the codename 'Frost', his handlers considered him an ambitious self-publicist who exaggerated his importance in Hollywood. But his extensive contacts in the film industry, US politics and the military made him a valuable asset. In 1943 Morros first came to the FBI's attention after being named as an NKVD source in Vasili Mironov's letter to J. Edgar Hoover (see Chapter Two) and placed under surveillance. When finally confronted by the FBI four years later he confessed his involvement with the NKVD and agreed to become a double agent. For the next nine years, under close FBI supervision, he fed false intel back to his handlers.

Around 1954, Morros' NKVD handler ordered him to use his contacts in the US military to find out how the Americans had dealt with the issue of tyre blowout on their new generation of supersonic jets, a problem which at the time was plaguing Soviet jets, like the MiG-19. 'He [his handler] explained to me that Russia had difficulty with supersonic planes, faster-than-sound planes ... when they land, they turn over and they lose the airplane,' Morros later testified.[1] The Americans had solved this problem by incorporating Orlon and Perlon, extremely strong synthetic fibres developed in the 1930s, into the tyres of their aircraft. The FBI gave Morros incorrect data on the formula of these fibres to pass on to his handler.

Arrows, Dragons and Mirages

The difficulties their intelligence operatives were experiencing in the US forced Moscow Centre to increasingly concentrate its efforts on countries they considered easier to penetrate.

One of these was Canada. During the 1950s, Avro Canada developed what was widely considered to be the most advanced interceptor in the world – and what would become the most controversial industrial project in Canadian history – the Avro Arrow.

The project had been instigated in April 1953, when the RCAF issued specification AIR 7-3, calling for a long-range supersonic interceptor with a two-man crew, capable of a cruising speed of Mach 1.5 and a combat radius of 300 nautical miles. Avro's proposal, the CF-105 Arrow, was accepted and the first of five prototypes unveiled on 4 October 1957, with flight testing beginning the following year. During these flight tests the Arrow proved to have world-beating performance, reaching Mach 1.9 in level flight.

Unsurprisingly, the Soviets followed the Arrow's development with keen interest. They were aided in this by a spy at the heart of the project. Known only by his codename 'Lind' (his true identity has never been publicly disclosed), he was an Irish-Canadian engineer with communist sympathies, who worked at Avro's Toronto plant. Lind supplied his handler Yevgeni Brik (codename 'Hart'), an 'illegal' living under the cover identity of an actual Canadian citizen, David Soboloff, with technical documents and photographs of the Arrow.[2] Brik's alcohol problem and messy private life, however, was to compromise the intelligence operation. After confessing to his married Canadian lover that he was actually a Soviet spy, Brik followed her advice and turned himself in to the RCMP in November 1953. Their Special Branch unit chose to run Brik as a double agent, feeding false or poor quality intelligence – including deliberately blurred photographs of the Arrow plans – back to Moscow Centre, in an operation known as Keystone.

Brik's double game was finally exposed to the KGB in July 1955, when James Morrison, a debt-ridden corporal in the RCMP, tipped off an official at the Soviet embassy in Ottawa that Brik was now working for the Canadians. His reward was 5,000 Canadian dollars. Brik, unaware he was now blown, returned to Moscow the following month to see his wife. Arrested by the KGB, he admitted under interrogation to working as a double agent and on 4 September 1956 was sentenced to fifteen years imprisonment.

Ironically, the Arrow itself would never enter service, falling victim not to Soviet espionage but to domestic politics. Concerned about the rising costs of the project, in February 1959 the Canadian prime minister, John Diefenbaker, announced the cancellation of this remarkable aircraft. In a highly controversial decision – which may or may not have been influenced by the Soviet espionage surrounding the project – all the prototypes and documentation relating to the Arrow were ordered destroyed.

Another advanced interceptor project of this period infiltrated by Soviet intelligence was the SAAB 35 *Draken* ('Dragon'). Although not a member of NATO and officially neutral in the Cold War, Sweden's geographical position would put the country in the frontline of any military confrontation between the Warsaw Pact and NATO, and so air defence was accorded a high priority by Stockholm. In the early 1950s SAAB began work on a new interceptor for the Swedish Air Force, the *Flygvapen*, to defend the country from Soviet bombers. This eventually emerged as the *Draken*, which featured a unique 'double delta' wing layout and was powered by a single Rolls-Royce Avon afterburning engine. The *Draken* became Europe's first supersonic interceptor, and began entering *Flygvapen* service in late 1959.

The GRU gained much intel on the *Draken* from a highly placed source in the *Flygvapen*. Colonel Stig Wennerström served as air attaché in the Swedish embassy in Moscow in 1940-41, and again in the late 1940s, during which time he is believed to have been recruited by the GRU. In 1952 he took up the same post in the Washington embassy, and in 1957 was appointed as a senior aide to the Swedish Minister of Defence. Known by the codename 'Eagle', besides information on the *Draken* Wennerström provided his Soviet handlers with details of US and British missile systems, Sweden's air defences, and the successor to the *Draken*, which was then in development, the SAAB *Viggen* ('Thunderbolt'). The net began to close in over Wennerström after the Americans were tipped off by Michal Goleniewski, an officer in the Polish SB intelligence service who defected to the West in 1961, that the Soviets had a high-ranking mole in the Swedish Air Force. Confirmation that Wennerström was the traitor came in 1963 when his housekeeper found rolls of microfilm containing classified documents hidden in his house and alerted the police. He was arrested on 20 June and sentenced to life imprisonment, later commuted to twenty years. Between 1957 and 1963, Wennerström is believed to have passed on more than 20,000 classified documents to the GRU.

One of the most successful European interceptors of the 1960s was the Dassault Mirage III, the pride of the post-war French aviation industry. Entering service with the *Armée de 'l'Air* in 1961, this delta-winged, supersonic aircraft was originally designed as an interceptor to shoot down Soviet bombers, although the Mirage proved itself a formidable dogfighter in the hands of Israeli pilots in the Middle East conflicts of the 1960s and '70s. Equipping the air forces of several NATO members and close US allies, the Mirage was inevitably of considerable interest to the Soviets, and in 1969 the KGB conceived a plan to entice a Lebanese Air Force pilot to defect with his Mirage.

Using an ex-Lebanese Air Force flight instructor discharged for misconduct, Hassan Badawi, as a go-between, KGB officers stationed at the Soviet embassy in Beirut hired LAF pilot Lieutenant Mahmoud Matar to deliver the Mirage, agreeing a reward of $2 million (with an upfront instalment of $200,000) and a new life in Switzerland to fly his aircraft to Baku in Azerbaijan during a routine training sortie.

Several meetings were held between Lieutenant Matar and the KGB officers in various apartments in Beirut to finalise the details of the plan. At the last of these, on 30 September 1969, the apartment at which Badawi, Matar and their KGB contacts, Vladimir Vasiliev and Colonel Aleksandr Komiakov, were meeting was raided by the Lebanese police. A gunfight broke out, and two Lebanese policeman and both KGB officers were wounded, Komiakov seriously. The KGB had been set up, Lt Matar having immediately reported the Soviet approach to his superiors, leading to a sting operation being organized by the Lebanese.

After angry diplomatic protests were lodged by Moscow, the Lebanese authorities quietly deported Vasiliev and Komiakov to the USSR on an Aeroflot flight, whilst both were still recovering from their wounds, on 4 October 1969.[3]

Peeking behind the Iron Curtain

One area of military aviation where the United States enjoyed a clear advantage over the Soviets was strategic reconnaissance. In the 1950s this advantage was most clearly embodied in the shape of the Lockheed U-2. Developed in great secrecy for the CIA, the U-2 could fly at altitudes over 75,000ft – far above the ceiling of any Soviet fighter then in service – and with its high-resolution Model A-2 cameras could provide detailed imagery from great heights. The U-2 made its maiden flight in August 1955 and carried out its first operational reconnaissance sortie over Soviet territory on 4 July 1956.

One of the primary missions of the U-2 in its first years of service was to establish if the so-called 'bomber gap' actually existed. This was a belief, perpetuated by SAC's belligerent Commander-in-Chief General Curtis LeMay, that the Soviet bomber fleet was far larger than SAC's. Photographs taken of Soviet Air Force bomber bases by U-2s on secret reconnaissance flights were able to prove that SAC's estimates of Soviet bomber strength had been grossly inflated and that the 'bomber gap' didn't exist.

The nearest Soviet equivalent to the U-2 was the Yak-25 RV (*Razvedchick Vysotnyj* – 'high-altitude reconnaissance'), introduced in 1959. But the Yak-25 fell far short of the American spy plane's maximum ceiling. Soviet efforts to build their own dedicated high-altitude reconnaissance platform received a boost thanks to one of the most infamous episodes of the Cold War. On 1 May 1960, a U-2 flown by Captain Francis Gary Powers was shot down by an S-75 surface-to-air missile while flying at an altitude of 70,000ft near Sverdlovsk, provoking a major diplomatic row between Moscow and Washington. The U-2 wreckage was recovered by the Soviets and it was subsequently decided that an attempt to reverse-engineer the aircraft should be made.

The task was handed to the Beriev OKB, which was better known for its seaplanes. With nothing but the shattered remnants of Powers' aircraft to work from, however, this would present an even greater engineering challenge than previous Soviet reverse-engineering projects, like the Tu-4 bomber. Designated the S-13, five examples were ordered. But progress was slow, and Khrushchev, increasingly convinced that the future of strategic reconnaissance lay in satellites rather than manned aircraft, cancelled the S-13 project in May 1962.

The political fallout from the U-2 shootdown incident encouraged the development of UAVS by the US military for strategic reconnaissance.

These had the obvious advantage that, should they be brought down over Soviet or Chinese territory, there was no pilot to be captured, interrogated and paraded at a show trial, as had been the case with Captain Powers.

The most sophisticated UAV project of the 1960s was the Lockheed D-21, work on which began in October 1962 under the codename 'Tagboard'. Sharing the titanium construction and much of the technology of the same company's A-12 stratospheric spy plane – later redesignated the SR-71 'Blackbird' – the D-21 was an air-launched drone powered by a Marquadt RJ43 ramjet engine, giving it a maximum speed of Mach 3.3 and service ceiling of 95,000ft. Range was 5,600km. The launch platform was the A-12, two of which were modified as carriers and redesignated M-21 (the 'M' standing for 'Mother', while the 'D' in D-21 denoted 'Daughter'). A Hycon HR 335 high-resolution camera was carried in a jettisonable module. The cost of the first twenty drones (of 38 eventually built) and the modifications to the two A-12s was put at $53 million.

On operational sorties, the D-21 would follow a preprogrammed route, photograph the target area and then release the module containing the camera over friendly territory for retrieval by either a JC-130 Hercules or a ship, should it land in water (for which flotation devices were fitted to the module). The drone itself would then self-destruct.

Testing began in March 1966, but did not go smoothly. On one of the test launches, both the D-21 drone and its M-21 carrier were destroyed, killing one of the M-21's crew. After this accident, improvements were made to the drone, resulting in the D-21B model, and it was decided to substitute the M-21 with the B-52H Superfortress as the launch aircraft. After several more unsuccessful launches, on 16 June 1968 the first satisfactory test flight was completed. Despite the technical problems revealed during the testing phase, the go-ahead was given for operational missions to be flown over China using the B-52H/D-21B combination, under the codename 'Senior Bowl'.

The first mission, to photograph the Lop Nur nuclear test site in the Xinjiang region in northwestern China (where the first Chinese nuclear bomb was detonated in 1964) was launched from Guam on 9 November 1969. Unfortunately, the drone's guidance system malfunctioned and it flew on into neighbouring Mongolia, where it crashed. Three more missions also failed due to technical issues. The high failure rate led to the cancellation of the D-21 programme on 23 July 1971.

However, the wreckage of the D-21 that crashed in Mongolia was recovered by the Soviets. Naturally, this cutting-edge reconnaissance platform was of considerable interest to Moscow and in March 1971 the Presidium of the Soviet Council of Ministers issued a directive to reverse-engineer the drone. Since it had established a subdivision in 1956 (known as Section K) to develop UAVs, the Tupolev OKB was selected to produce the Soviet copy, which was named *Voron* ('Raven').[4] The RD-012 ramjet engine was chosen to replace the original's RJ43 unit, which would give the *Voron* a similar maximum speed to the D-21, while the Tu-95K was to be used as the launch platform.

The *Voron* project was overseen by the OKB's chief designer Alexei Tupolev, son of the company's founder. But the technical challenges of reverse-engineering the D-21 proved insurmountable for the Soviet engineers, and after several years of development work the project was abandoned without any Soviet clones having been produced.[5]

Directorate T

The growing importance to the Soviet economy and military of S & T intelligence led to the creation in 1963 of a new department within the KGB's FCD, the Directorate of Scientific and Technical Intelligence (known simply as Directorate T), to oversee and coordinate the activities of Line X officers in the West. Directorate T had responsibility for fulfilling the requests for information on Western military technology, which came primarily from the Military-Industrial Commission, the VPK. The first head of Directorate T was the KGB's long-serving Chief of Scientific Intelligence, Colonel Leonid Krasnikov, but after his retirement in 1964, Colonel Mikhail Lopatin was put in charge.

Some of Directorate T's greatest successes in the 1960s were in Western Europe, with Line X officers based at the London residency proving especially effective at recruiting sources. One of the most important of these was an aeronautical engineer employed by British European Airways, codenamed 'Ace'. Recruited in 1967, 'Ace' was the KGB's main source of intelligence in the UK on the Anglo-French Concorde supersonic airliner. Though working in the civil aviation sector, the intel he passed on to his handlers also had considerable military value; 'Ace' supplied information on Concorde's Rolls-Royce Olympus 593 engines, earlier versions of which powered the Avro Vulcan and BAC TSR 2 tactical strike and reconnaissance aircraft.[6] He also gave the KGB details of the Rolls-Royce Spey turbojet engine, which besides being used in airliners like the BAC 1-11 and Hawker Siddeley Trident also powered the Blackburn Buccaneer strike aircraft and British Aerospace Nimrod maritime patroller.

'Ace' continued working for the Soviets until his death in 1982, supplying around 500 volumes of documents. MI5 only discovered the existence of this spy after the defection to the West of Vasili Mitrokhin, a senior KGB archivist, in 1992. An MI5 report estimated that 'Ace' saved the Soviets millions of roubles in research and development costs.[7]

Another agent – albeit one of far less importance – recruited around the same time was Clive Bland. A 20-year-old photoprinter employed by the Ministry of Technology at RAE Farnborough, Bland handed over 11 documents relating to aircraft and guided missile tests to the Soviet embassy in London. Arrested in 1968, he pleaded guilty to breaching the Official Secrets Act. While his defence lawyer insisted Bland was no professional spy but merely 'a Walter Mitty character who dreams himself into something he is not,' prosecutor Peter Barnes told the court that the Ministry of Technology 'regards disclosure [of the documents] as a serious matter.'[8] On this occasion, the court showed leniency, merely fining Bland £50 for handing over the documents.

Besides the KGB, the GRU was also active recruiting British aviation spies. One of the prize assets the GRU ran in London in the 1960s was Frank Bossard. Born in 1912, Bossard joined the RAF in 1939, despite having a previous conviction for cheque fraud. He gained promotion to Flight Lieutenant (apparently thanks to a falsified CV), and served in the Mediterranean and Middle East. After the war, he joined the Ministry of Aviation, serving with its Scientific and Technical Intelligence Branch as a signals officer, stationed in West Germany. In 1956 he was recruited by SIS. Based at the British embassy in Bonn, his duties included debriefing Soviet aircraft engineers who had come over to the West. In 1961 he returned to the Ministry of Aviation and was posted to London. Bossard's alcohol and money problems made him vulnerable to recruitment by the Soviets, and he began passing on photographs of documents relating to guided missiles and radar to the GRU, receiving £5,000 for the material he supplied. A GRU double agent working for the CIA, Dmitri Polyakov, identified him as a Soviet asset and he was arrested at the Ivanhoe Hotel in London on 12 March 1965 while photographing secret documents. After pleading guilty at his trial in May 1965, he was sentenced to 21 years. Released in 1975, he changed his name to Frank Clifton and died in 2001.

The intelligence services of Moscow's Warsaw Pact allies also enjoyed some success in the UK. In 1961, the Czech StB recruited Nicholas Praeger (codename 'Marconi'), the son of a Czech clerk at the British embassy in Prague during the Second World War. In 1949, Praeger emigrated to the UK and, after being granted citizenship, joined the RAF as a radar technician, serving with the

Above: The Polikarpov R-1, an unlicensed copy of the British Airco DH9.A – several examples of which fell into Bolshevik hands during the Russian Civil War – was the first Soviet military aircraft based on a Western design.

Below: The United States provided the USSR with thousands of aircraft during the Second World War through the Lend-Lease scheme, including these B-25 Mitchell bombers awaiting delivery to the Red Air Force. The Americans took the precaution of removing the more sensitive equipment, like the top-secret Norden bombsight, before handing over the aircraft. (US Air Force)

Above: The unorthodox Consolidated Vultee XP-81, a long-range fighter powered by both a jet and a turboprop engine, was one of several experimental wartime aircraft Soviet spies in the US acquired intelligence on. (US Air Force)

Below: One of the most ambitious reverse-engineering projects ever undertaken, the Tupolev Tu-4 was copied from an American B-29 Superfortress that crash-landed on Soviet territory in 1944, becoming the USSR's first atomic bomber. (Courtesy of Alf van Beem)

Above: America's first jet-powered aircraft, the Bell P-59 Airacomet, was a target for Soviet intelligence's Project *Vozdukh*, an operation to steal British and American gas-turbine engine technology. (US Air Force)

Below: The Klimov VK-1 gas-turbine engine, which powered the MiG-15 and other early Soviet jets, was a pirated copy of the Rolls-Royce Nene, 25 of which were sold to the USSR by the British Government in 1946-47. (US Air Force)

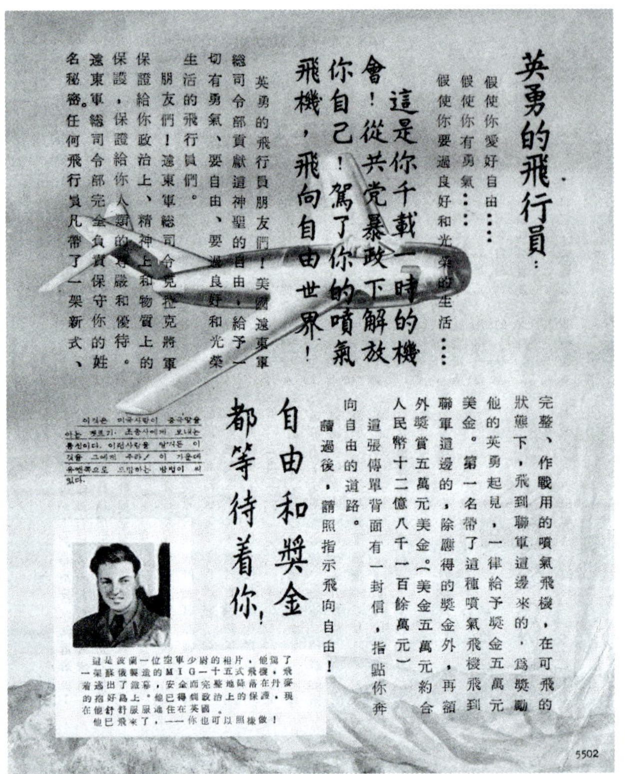

Thousands of leaflets like this were dropped over North Korea as part of Operation Moolah, offering a $100,000 reward and political asylum to the first communist pilot to defect to South Korea in a MiG-15. The photograph is of Franciszek Jarecki, a Polish MiG pilot who sought refuge in Denmark in March 1953. (US Air Force)

Lt No Kum-sok of the North Korean People's Air Force – seen here in his flight suit – took up the American offer, flying his MiG-15 to Kimpo air base on 21 September 1953, shortly after the end of the Korean war. (US Air Force)

Above: No Kum-sok's MiG-15, now sporting USAF markings, under armed guard at Okinawa, where it underwent initial testing before being shipped to Wright-Patterson Air Force Base in Ohio for more extensive flight testing by specialists from the Air Technical Intelligence Center. (US Air Force)

Below: The appearance of the MiG-25 in the mid-1960s sparked considerable alarm in NATO circles and much debate about its capabilities, until examination of Soviet defector Lt Viktor Belenko's 'Foxbat' by US analysts stripped away much of the mystique surrounding this powerful but relatively crude interceptor. This Iraqi Air Force MiG-25 was discovered buried at the Al Taqaddum air base by US troops during Operation Iraqi Freedom in 2003. (US Air Force/Master Sgt. T. Collins)

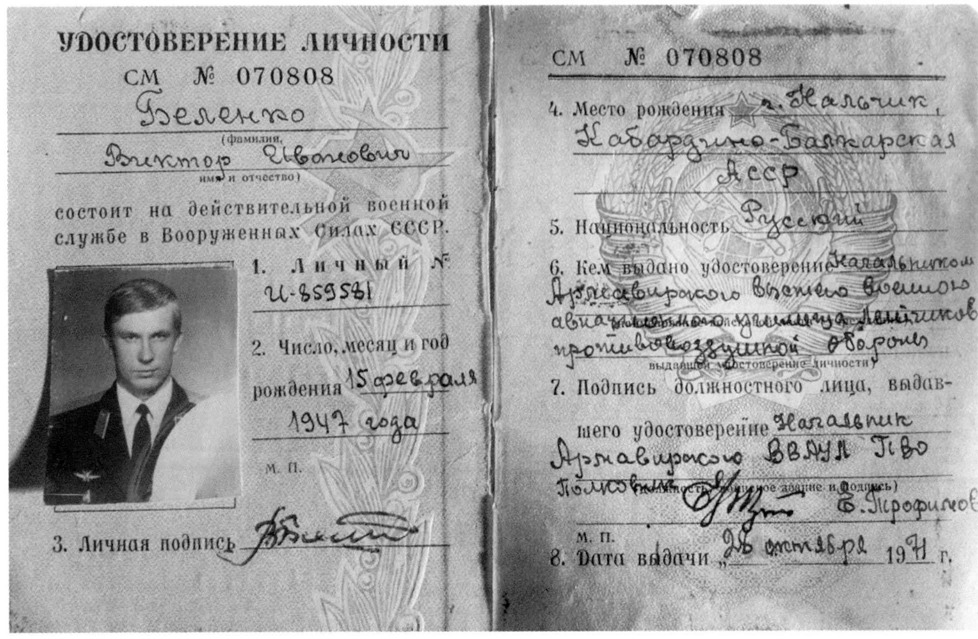

Above: Lt Belenko's military ID card. Disillusioned with communism, the pilot defected to Japan in his MiG-25 in September 1976, handing the West an 'intelligence bonanza' in the words of the then Director of the CIA, George H.W. Bush. (CIA Museum)

Below: Captain Francis Gary Powers (right) and Lockheed chief Kelly Johnson with a U-2 spy plane. After Powers was shot down over Soviet territory on a reconnaissance mission in a U-2 in 1960 the Soviets used the wreckage as the basis for a reverse-engineered copy. The project was abandoned before any copies were completed. (US Air Force)

Another Lockheed reconnaissance platform the Soviets attempted without success to reverse-engineer was the highly advanced D-21 Unmanned Aerial Vehicle, one of which crashed in Mongolia in 1969. (US Air Force)

In the 1960s, the KGB and the Czech StB received information on the Olympus engine and 'Blue Diver' ECM equipment used on the RAF's Vulcan nuclear bomber from spies in the British aviation industry. (© Crown copyright)

Left: Meir Amit, Director of the Israeli Mossad intelligence service, masterminded Operation Diamond, an elaborate mission to orchestrate the defection to Israel of an Iraqi Air Force pilot in his MiG-21 in 1966. (© National Photo Collection of Israel/Yaacov Saar)

Below: After being exhaustively tested by the Israelis, the Iraqi MiG-21 – seen here in USAF markings – was handed over to the United States and put through an equally rigorous programme of technical evaluation and flight testing by the Foreign Technology Division at Groom Lake in Nevada, under the codename Project Have Doughnut. (US Air Force)

Above: The CIA secretly negotiated a deal with Egypt for several of their MiG-23 'Flogger' fighters in the 1970s. The 'Floggers' were tested by the USAF's secretive 4477th Test and Evaluation Flight – nicknamed the 'Red Eagles'. (US Air Force/ Ken LaRock)

Below: An intelligence windfall unexpectedly came NATO's way in April 1966 when one of the Soviet Air Force's most advanced interceptors, the Yak-28P 'Firebar', crashed in the British zone of West Berlin. British military personnel were quickly on the scene to secure items of interest from the wreckage. (US National Archives and Records Administration)

Above: The most important aviation prize captured by the Soviets during the Korean War was an almost intact North American F-86E Sabre. These F-86Es are pictured at a South Korean air base in 1952. (US Air Force)

Below: Pictured here mounted on a MiG-21, the K-13 (known to NATO as the AA-2 'Atoll') was a reverse-engineered Soviet copy of the AIM-9B Sidewinder air-to-air missile, one of which was passed on to Moscow by the Chinese in 1958 after becoming lodged in a J-5 fighter during combat with the Taiwanese. (US Air Force/Ken LaRock)

Above: The wreckage of US aircraft shot down during the Vietnam War offered both the Soviets and the Chinese plentiful opportunities to examine the latest American aviation technology. This F-105 Thunderchief has been fatally damaged by an S-75 missile over North Vietnam. (US Air Force)

Below: One of the most valuable items of hardware recovered from a downed US warplane in Vietnam was the QRC-160 pod, which jammed the S-75 SAM system's radar. The capture of an intact QRC-160 pod in February 1968 allowed Soviet engineers to devise countermeasures to the jammer. (US Air Force)

Above: The Soviet Army's Mi-24 'Hind' was a heavily-armed assault helicopter, much feared by NATO. The armour plating and other components from a 'Hind' shot down by the *Mujahideen* during the Soviet-Afghan War were extracted from Afghanistan in a covert British operation. (UK MOD © Crown copyright)

Below: In 1988, the Americans acquired an intact Libyan 'Hind' (seen here slung beneath a US Army Chinook) in a mission codenamed Operation Mount Hope III. (US Department of Defense)

Above: The Grumman F-14 Tomcat and AIM-54 Phoenix long-range air-to-air missile were a potent combination when introduced into US Navy service in the 1970s. The crash of an F-14 off the Orkneys in 1976 prompted a massive recovery effort by the US Navy to prevent the wreckage being retrieved by the Soviets. (US National Archives and Records Administration)

Below: The plans for the Panavia Tornado, a multinational European strike aircraft, were passed on to the KGB by a highly-placed source in the West German aerospace giant MBB. (UK MOD © Crown Copyright)

The introduction by the USAF of stealth aircraft in the 1980s, like the Lockheed F-117 Nighthawk, promised to revolutionize aerial warfare. Lagging behind the US in the development of this technology, the Soviets turned to espionage to uncover the secrets of stealth. (US Air Force/ Ken LaRock)

The second stealth aircraft to enter USAF service, the Northrop Grumman B-2 Spirit, became a target for both the Soviet and Chinese intelligence services. In 2010 an engineer who worked on the B-2 programme was convicted of selling secrets of the bomber to China. (US Air Force/Master Sgt. Kenneth Norman)

In the 1980s, the MiG-31 'Foxhound' equipped with the Zaslon look-down/ shoot-down radar was the Soviet Air Force's most formidable interceptor. The CIA uncovered full details of the Zaslon through their spy in the Soviet radar industry, Adolf Tolkachev. (US Department of Defense)

Above: The fall of the Iron Curtain opened up new opportunities for the West to examine – and even purchase – Soviet warplanes from former Warsaw Pact air forces, such as the MiG-29 'Fulcrum'. This is one of 21 MiG-29s bought by the US from Moldova in 1997. (US Air Force/Ken LaRock)

Below: The cockpit canopy of the F-117 shot down over Serbia during Operation Allied Force in 1999, on display at the Museum of Aviation in Belgrade. Pieces of wreckage from the Nighthawk were allegedly handed over to the Russians and Chinese to assist their own stealth aircraft programmes. (Courtesy of Petar Milošević)

Above: The Russo-Ukrainian War has seen both Russia and NATO deploy their latest weaponry. The Anglo-French Storm Shadow air-launched missile provided to the Ukrainians, pictured here carried on an RAF Tornado, has proven highly effective on the battlefield. An almost intact Storm Shadow was captured by Russian forces in 2024. (© Crown copyright)

Below: The Lockheed F-35 Joint Strike Fighter, operated by several NATO air forces, has been the focus of intensive Chinese espionage. The similarities between the F-35 and the Shenyang J-35 have led to much speculation in the West that the Chinese copied the design. (UK MOD © Crown Copyright)

Radar Development Unit at RAF Finningley where, according to court documents, he was 'one of a very small team engaged on highly secret and important work.' Praeger passed on photographs he took of documents on ECM equipment fitted to the RAF's 'V' bomber force, codenamed 'Blue Diver' and 'Red Steer', to his contact at the Czech embassy in London, known to him as 'Malek'. After leaving the RAF to join the English Electric Company, maker of the Canberra reconnaissance aircraft and Lightning fighter, he continued passing secrets to the Czechs. He was finally unmasked as a spy when StB officer Josef Frolik defected to the US in 1969 and exposed the Czechs' intelligence operations in the West. Arrested by Special Branch in 1971, he was found guilty, sentenced to 12 years (later reduced to six) and stripped of his British citizenship.

By the late 1960s the Federal Republic of West Germany was assuming ever greater importance in S & T intelligence collection, and would soon become the Soviets' main target in Western Europe, accounting for over ten per cent of all S & T intelligence collected by 1980. Aiding the KGB in this task was East Germany's HVA, whose equivalent of Directorate T was the SWT (*Sektor Wissenschaft und Technik* – Sector for Science and Technology), established in 1971. The SWT proved to be one of the most successful of the Soviets' satellite spy agencies; by 1988, it was running over 300 informants in West German companies, universities and research institutes.

The SWT's most important asset, and reputedly its highest paid, was Wolf-Heinrich Prellwitz. An official in the FRG's Ministry of Defence, Prellwitz was recruited by an HVA officer in 1968, who posed as a business contact for French arms industry lobbyists. Given the codename '*Rödel*' (meaning 'Toboggan'), Prellwitz supplied his handlers with tens of thousands of photographs of classified documents, including ones on Luftwaffe strategy in the event of war and projects under development at West German aircraft manufacturers, such as the Panavia Tornado strike aircraft and the

Franco-German Eurocopter Tiger gunship. Motivated by money, he was paid 820,000 deutschemarks over the twenty years he was active, though his modest lifestyle ensured he aroused no suspicion. After German reunification, a mole-hunt was launched after an ex-HVA officer revealed to German military counterintelligence that they had a highly placed source in the Ministry of Defence. '*Rödel*' was finally identified and arrested in April 1991 and on 21 May 1992 was sentenced to 10 years at a court in Dusseldorf for treason.

The SWT had particular success penetrating West Germany's largest aerospace firm, Messerschmitt-Bonow-Bölkow (MBB). One of the longest-serving of their assets was Dieter Feuerstein (codename 'Petermann'). After studying aerospace engineering in West Berlin, he worked as an engineer at MBB and was recruited by the HVA in 1974. Among the intel he provided were the designs of the company's submission for the 'Tactical Combat Aircraft 90' competition, which eventually evolved into the Eurofighter Typhoon. Like Prellwitz, Feuerstein was only finally exposed after German reunification. In 1992, he was sentenced to eight years by the Bavarian Supreme Court.

Swiping a Sidewinder

It was a gang of agents working on behalf of the Soviet GRU, however, that carried out what must surely rank as the most brazen act of military technology theft in the FRG during the Cold War.

It took place on a foggy night in October 1967, when three men broke into the Neuberg Luftwaffe base in Bavaria, cutting through the perimeter fence and gaining access to the ammunition magazine. Loading a Sidewinder missile into a wheelbarrow, they snuck back out of the base to their waiting Mercedes getaway car without raising the alarm. The heist then took a comical turn when the thieves realized that the near three-metre-long missile wouldn't fit in the vehicle, forcing them to break one of the rear windows to accommodate it and concealing the protruding section of missile with a blanket, before driving off.

The thieves were Josef Linowski, a Polish mechanic, Wolf-Diethardt Knoppe, an F-104 Starfighter pilot in the West German Luftwaffe stationed at the Neuberg base, and Manfred Ramminger, the leader of the group, a 37-year-old architect from Krefeld with a taste for the high life. A year earlier Ramminger, in need of cash, offered to acquire West German military hardware for the GRU. He proved his abilities in April 1967, when he succeeded in stealing aircraft navigation equipment from the Neuberg air base. He then offered to provide the Soviets with an AIM-9B FGW.2, an upgraded, licence-built West German version of the standard 'B' model Sidewinder, with an improved infrared seeker head. This would be quite a prize for the Soviets, as their standard heatseeking AAM, the Vympel K-13 – based on an earlier version of the Sidewinder – had rather limited capabilities, and the GRU gratefully took up his offer.

After making his getaway from the air base, Ramminger disassembled the missile, crated up the components, and simply mailed it to Moscow as commercial air freight. The architect received 92,000 deutsche marks (about $8,500) for his troubles from the GRU. But he had little opportunity to enjoy his reward, for he, Linowski and Knoppe were arrested a few months later after Linowski drunkenly boasted about the heist in a bar.[9] Found guilty in a Dusseldorf court of espionage and grand larceny in October 1970, Ramminger and Linowski were each sentenced to four years, while Knoppe received three years and three months. Ramminger, however, served only a few months before being exchanged in a prisoner swap. Some of the technology gleaned from the stolen Sidewinder by Soviet engineers found its way into the improved K-13M variant.

Chapter Eight

The Coldest Years Part II
Western Intelligence Operations 1953 – 1969

On a November night in 1953, a USAF Douglas C-124 heavy-lift transporter of the 62nd Troop Carrier Wing, based at Larson AFB in Washington state, landed at a military airfield somewhere in the Balkans. Under conditions of strict secrecy, waiting air force personnel loaded several heavy crates onto the C-124 and it took off again, heading back across the Atlantic and landing at Wright-Patterson AFB, where the cargo was unloaded and reassembled in a hangar by personnel of ATIC. When the technicians were finished, they had a complete Yak-23.

Assigned the reporting name 'Flora' by NATO, the Yak-23 was a straight-wing jet fighter, powered by the RD-500 turbojet (Soviet copy of the Rolls-Royce Derwent V). First flying in July 1947, the performance of the Yak-23 was generally inferior to that of the rival MiG-15 and so production was limited, amounting to 319 examples (including three prototypes), as was its service life with the VVS, being largely replaced with the MiG-15 by 1952. Nevertheless, the Yak-23 remained in service with the air forces of several Soviet bloc countries, including Czechoslovakia, Romania, Poland and Bulgaria, until the late 1950s and was therefore of some interest to American intelligence.

The top-secret mission to smuggle the Yak-23 to the States was known as Project Alpha, and the Balkan country which handed over this prize to the Americans was Yugoslavia. Although a communist country, relations between Yugoslavia and the USSR were extremely strained at the time, due to Yugoslav leader Marshal Josip Tito pursuing his own independent policies, angering Stalin to the extent that he ordered Tito's assassination in early 1953. Tito was therefore keen to forge a stronger relationship with the US. When Romanian pilot Mihail Diaconu defected to Yugoslavia in a Yak-23 in June 1953 and claimed political asylum, he saw an ideal opportunity to achieve this, by offering CIA officers stationed at the US embassy in Belgrade the loan of Diaconu's fighter.

At Wright-Patterson, the Yak-23 made its first flight on 4 November, with test pilot Captain Tom Collins at the controls. Between then and 25 November, seven further flights were carried out. Inquisitive USAF personnel stationed at Wright-Patterson who weren't privy to Project Alpha were told that the Yak-23 – which had temporary USAF markings applied – was an experimental Bell X-5 aircraft. In their report, the test pilots described the Yak's rate of climb and acceleration as 'excellent', but highlighted its lack of cockpit pressurization and poor directional stability above 325 knots as serious shortcomings. In accordance with the deal negotiated with Belgrade, the Yak-23 was then dismantled and flown back to Yugoslavia on another C-124.

Project Alpha remained a closely guarded secret until after the Cold War, when limited information about the successful operation was finally declassified.

The success of Project Alpha, along with Operation Moolah two months earlier, was a much-needed boost for US intelligence at a time when their S & T acquisition efforts were coming in for high-level criticism within the Pentagon. In July 1953 the National Security Council's Intelligence Advisory Committee reported that 'the flow of information of a scientific and technical nature from conventional sources is becoming increasingly inadequate' and that in particular 'intelligence on weapons and equipments pertaining to the Soviet air offensive and defensive capabilities remains generally inadequate.'[1]

Despite the Yak-23 coup, by 1955 the SOVMAT Program was not living up to expectations, as an internal memo from the Assistant Director of the CIA's Office of Research and Reports (ORR) Otto Guthe frankly admitted:

> The development of the Sovmat Program has not been, for the most part, as was initially anticipated. While collection of the majority of Sovmat type items is a matter of opportunity and consequently can be geared only to a long-range research program, this Office has not been completely satisfied with respect to the length of time involved, the cost and the responsiveness of exploitation reports for these items.[2]

By the end of that year, however, the ORR was able to report that significant progress had been made, including a twenty per cent increase in the amount of SOVMAT items collected compared to the previous year. The ORR noted, however, a growing trend for agencies to request military hardware that was virtually unobtainable, including radar systems and guided missiles, and urged requestors to 'provide extremely good justification for acquisitions of this kind to permit establishment of complex operations and the offering of substantial financial inducements necessary to try to get the desired items.'[3]

Over the next few years there was a steady improvement in the performance of the CIA's SOVMAT Program. In 1960 a progress report revealed that, 'Further development of sources and methods has resulted in the production of more intelligence information from fewer, but more significant objects.' Improved liaison with other agencies had also 'led to greater efficiency of collection activities.'[4]

Among the more notable successes achieved by the CIA's SOVMAT collection teams in the 1960s was the acquisition of a Mil Mi-8. A general-purpose utility helicopter that could carry up to 24 troops, the Mi-8 made its public debut at the 1961 Tushino air show. Allocated the NATO reporting name 'Hip', over 8,000 were produced and it became the Warsaw Pact's standard general-purpose helicopter, as well as serving with many Soviet client states. Highly versatile, the 'Hip' was also converted for ASW, ECM and minesweeping duties, while an assault helicopter version was armed with a KV-4 12.7mm machine-gun in the nose and rocket pods and anti-tank guided missiles mounted on pylons.

In an operation known as Project Chuck Hole, the CIA acquired an Mi-8 from an undisclosed country in May 1966, the cost of this operation amounting to $643,000. But there was disagreement between the US Army and the USAF over who should analyze the helicopter. The Army insisted it should be sent to their base at Fort Benning, Georgia, while the director of the Defense Intelligence Agency Lieutenant General Joseph Carroll argued that evaluation of the 'Hip' be undertaken at Holloman AFB in New Mexico, due to 'the proximity of the White Sands facilities for radar and infrared testing,' as well as the site having 'the necessary security requirements.' The wrangling eventually ended in the Air Force's favour.[5]

The hunt for a 'Fishbed'

In the 1960s, there was one Soviet fighter above all others that the West wanted to examine in detail – the MiG-21 'Fishbed'. The MiG-21 story began in 1953, when the NII issued a specification for a supersonic point-defence interceptor capable of destroying the new generation of high-performance bombers such as the Boeing B-47 Stratojet and Convair B-58 Hustler which were being introduced into SAC service, and the 'V' bombers of RAF Bomber Command. The MiG OKB responded with a proposal for a lightweight, single-seat fighter with a small delta-wing and single afterburning Tumansky RD-11 engine. Initially designated the Ye-2, it eventually evolved into the MiG-21, which began entering VVS service from 1959. Although its armament was relatively light, consisting of a single GSh 23mm cannon and a pair of K–13 AAMs, the MiG-21's impressive maximum speed of 1,350 mph and rapid rate of climb ensured it was regarded as a potent threat by NATO. Over 12,000 were built in the USSR, Czechoslovakia and China (as the Chengdu J-7), serving with more than 40 communist and Soviet-allied countries, and it became the backbone of the Warsaw Pact's interceptor force for decades.

The MiG-21 enjoyed a particularly fearsome reputation in the Vietnam War. US reconnaissance confirmed the arrival of the first MiG-21s on a North Vietnamese airfield in December 1965, and they began operations in April 1966. VPAF MiG-21s claimed their first victim, an F-105D Thunderchief, on 7 July 1966. Even the Americans' most advanced fighter, the mighty F-4 Phantom, struggled against the less technically sophisticated but more nimble MiG during the Vietnam War. The legendary USAF ace Colonel Robin Olds judged the MiG-21, at altitude, superior to the F-4.[6] Acquiring an intact MiG-21 to learn what made it tick therefore became a prime objective for the CIA.

The man the Agency chose to fulfil this ambitious goal was Guenter Laudahn. A 30-year-old electrical engineer from Potsdam, Laudahn had fled to West Germany in 1962 and was recruited by the CIA at the Marienfelde reception centre for East German refugees. In April 1966 he was sent back into the GDR with false papers and orders to make contact with Klaus Dieter Junge, the brother of a MiG-21 pilot in the LSK, whom the CIA had been led to believe could be encouraged to defect in his aircraft, for a price. Helping Laudahn in his high-risk mission behind the Iron Curtain were two other East German refugees: 22-year-old waiter Hans-Juergen Hanke and 24-year-old lathe operator Werner Baecker, a member of an underground organization that assisted East Germans escape to the West.

Following the instructions given to him by his CIA handlers, Laudahn met with Junge in an East Berlin park to negotiate the terms of the two brothers' defection. 'I told him that the more equipment the MiG-21 jet fighter contained, the higher the price would be,' Laudahn recalled. 'I was told [by the CIA] landing plans would be supplied, East German radar devices would be made useless, both brothers would be looked after once they were in West Germany, and all their relatives would be got out of East Berlin.'[7]

But Junge had a change of heart and secretly informed the Stasi of the CIA approach. When Laudahn next travelled to East Berlin on 30 May to deliver a message in code to Junge, giving his brother final instructions for the flight, Stasi officers who were lying in wait grabbed him. Hanke and Baecker were also apprehended by the Stasi. Following a five-day trial at the GDR Supreme Court in East Berlin in August 1966, all three were found guilty of espionage. Laudahn was given life imprisonment, while Hanke and Baecker received sentences of ten and six years, respectively. After spending several years in solitary confinement, Laudahn was exchanged in a spy swap at the Glienicke Bridge (famously known during the Cold War as the 'Bridge of Spies').

Just a few months after this disastrous CIA mission, a very similar operation masterminded by the Israeli intelligence service Mossad succeeded where the Americans had failed.

Operation Diamond

By the early 1960s, the Soviets had begun exports of the MiG-21 to their client states in the Middle East. Before long, the air forces of Israel's Arab neighbours, including Egypt, Syria and Iraq, were all operating the type. As all these countries were sworn enemies of Israel, it was inevitable that Israeli fighter pilots would soon encounter the MiG-21 in combat.

In 1965, the head of Mossad Meir Amit held a meeting with General Ezer Weizman, chief of the Israeli Air Force, at which he asked him if there was anything his service could do for the IAF. Without hesitation, the General replied: 'We need a MiG-21.'[8] Amit knew this would be a very tall order. Two previous attempts by Mossad to encourage the defection of a MiG-21 pilot had both ended in failure. The first was in 1963, when a team of Mossad operatives offered an Egyptian Air Force pilot, Adib Hanna, a large cash sum to steal a MiG-21 and fly it to an Israeli air base. But Hanna alerted his superiors to the bribery attempt and the leader of the Mossad team was arrested by Egyptian police. Undeterred, soon after, Mossad again tried to bribe an Arab pilot into defecting in a MiG-21, this time in Iraq. The pilot they targeted, however, turned down their offer. To secure his silence, the Mossad team allegedly beat him up.

With General Weizman impressing upon Amit the importance of securing a MiG-21, the Mossad chief agreed to make another, all-out effort. Responsibility for fulfilling this task was handed to Rehavia Vardi, an experienced Mossad officer who had been involved in the earlier unsuccessful attempts to secure a MiG. Vardi instructed his contacts throughout the Middle East to identify any Arab MiG-21 pilot who might be willing to defect. Within a few weeks, he was tipped off about a promising possible candidate by one of these contacts, an Iraqi Jew named Yossef Shemesh.

Munir Redfa was a 32-year-old captain in a MiG-21 'Fishbed-E' squadron of the Iraqi Air Force. He was also one of the few Christians in the Iraqi military. Religious discrimination ensured he was denied the promotions he felt his skills as a pilot merited, and bred in the ambitious Redfa a deep resentment. Added to this was a growing disillusionment with the regime he served, after he was ordered to bomb rebel Kurdish villages full of women and children. Shemesh, who was having an affair with the sister of Redfa's wife Betty, was ordered to befriend the pilot. His cultivation of Redfa was a slow and painstaking process, and it was almost a year before he felt he had sufficiently earned the pilot's trust to move on to the next stage of the operation, which was called 'Yahalom' (Hebrew for 'Diamond', after the codename Mossad allotted to Redfa). Finally, Shemesh persuaded Redfa to travel to Athens with Betty, on the pretext that she required life-saving medical treatment unavailable in Iraq, to meet with a friend.

In Greece, Redfa was introduced to Colonel Ze'ev Liron, head of the IAF's Intelligence Division, who told the Iraqi he was a Polish pilot who did work for an underground anti-communist movement. After striking up a friendship, Liron brought Redfa to a small Greek island where he revealed he was actually an Israeli and asked if he would be willing to defect to Israel in a MiG-21. Redfa decided to sleep on the matter. The next day the Iraqi pilot gave Liron his answer – he would do it. Their conversation then turned to Redfa's reward. The Iraqi accepted Liron's offer of $300,000 (Meir Amit had authorized him to offer as much as $600,000), along with asylum in Israel for himself and his family.

They then flew to Rome, where another IAF intelligence officer, Yehuda Porat, met with Redfa to work out the details of his defection flight. Porat was impressed by the Iraqi, describing him as a 'man of honour'. Amit, who secretly observed Redfa during one of Porat's meetings with him in a Rome café, was similarly impressed with the 'short, sturdy, and broad-shouldered man with a serious face.'⁹ Convinced they had chosen the right man, Amit gave the Mossad team handling Redfa the green light to proceed with Operation Diamond.

A more intensive briefing took place in a secret Mossad facility in Tel Aviv to prepare the pilot for his dangerous mission. Redfa was told that he was to fly from Rasheed air base, 11km southeast of Baghdad, to Hatzor in southern Israel. As the MiG-21 lacked the range to complete this nearly 500-mile-long flight, he would have to ensure his aircraft was fitted with long-range drop tanks. Once in Israeli airspace, he was told, he would be met by a pair of IAF Mirage fighters, which would escort him to Hatzor. With all the details worked out, Redfa then returned to Iraq via Athens. The pilot, however, risked blowing the operation when he decided to sell off all his worldly belongings ahead of his defection. 'I was scared to death,' Amit later confessed, 'that the Iraqi *Mukhabarat* [secret police] would find out about the sale, would interrogate Redfa, arrest him, and the entire operation would collapse.'[10] Fortunately, the sale failed to attract the attention of the much-feared *Mukhabarat*.

Mossad's next step was to quietly exfiltrate Redfa's family. They arranged for Betty and the couple's two young children to be flown to Amsterdam and then taken to Paris. Redfa had not confided to his wife his intention to defect, fearing a negative reaction; she only learned of her husband's plans when she was taken to an apartment in Paris, where rather than being reunited with her husband as she had expected, Colonel Liron told her of Munir's deal to fly his aircraft to Israel. Betty did not take this news well. According to Liron, she burst into tears, called her husband a traitor and threatened to expose the scheme to officials at the Iraqi embassy in Paris. 'She didn't stop screaming and crying all night long,' Liron recalled. 'I tried to calm her down; I told her that if she wanted to see him, she had to come to Israel with me. She realized she had no other choice.'[11]

Meanwhile, in Iraq, Redfa heard the popular Arabic folk song *Marhabtayn Marhabtayn* ('Welcome Welcome') on Israeli radio. This was the agreed signal for Redfa to make his flight. He took off from Rasheed on the morning of Sunday, 14 August 1966, on a training sortie. But soon after take-off his cockpit began filling with smoke. Captain Redfa abandoned the flight and landed back at Rasheed, where groundcrew discovered that the smoke was due to a blown fuse in the electrical system.

The next attempt took place two days later, on the 16th. At 6.55 am, he again took off from Rasheed air base. Following the flight path worked out in advance with his Mossad handlers, this time the flight went off smoothly, although he was pursued by Hawker Hunters of the Royal Jordanian Air Force while flying over Jordanian airspace (the Hunters were unable to keep pace with the MiG and broke off the pursuit). After crossing into Israeli airspace, Captain Redfa's MiG-21 was joined by his escort of Mirages, flown by Lieutenant Colonel Shamuel Shefer and Ran Ronen, CO of the IAF's No 119 Squadron. To demonstrate his non-hostile intent, Redfa tilted his wings and gave the Israeli pilots the thumbs-up sign. At eight am the MiG and the two Mirages landed safely at Hatzor.

The Iraqi defector was warmly greeted by the base commander and, after giving a press conference for the Israeli media, was taken to a house in Herzliya, near Tel Aviv, where he was reunited with his wife and children. But Redfa had difficulty adjusting to his new life in Israel. 'After his arrival,' revealed a Mossad officer, 'Munir had a very hard, miserable, and sad life.'[12] His defection caused serious splits amongst his extended family, with his brother-in-law, an officer in the Iraqi Army, denouncing him as a traitor. Redfa found work as a pilot in the Israeli oil industry, but his wife struggled with their new circumstances and eventually they left Israel to settle in an unnamed Western country. Despite being given new identities, the family continued to live in fear of the vengeful Iraqi security services.

Redfa's MiG-21 (which was given the new serial number '007') was subjected to an intensive programme of testing in Israel. The IAF's chief test pilot, the highly experienced Danny Shapira, was chosen to make the first test flights. To prepare for his task, he met with Redfa to learn as much as he could about the aircraft. The Iraqi had also helpfully brought along a copy of the operator's manual with him on his defection flight.

On 13 September 1966, Shapira took off in the MiG from Hatzor air base, watched anxiously by Redfa and a number of senior IAF officers, including General Weizman, who had pleaded with Shapira beforehand to bring the prized aircraft back in one piece. This first test flight went without a hitch. Over the following days, Shapira completed several more test flights, which included pitting the MiG-21 against IAF Mirages in mock dogfights.

The knowledge gained from these test flights was to prove invaluable to the IAF, giving the Israeli pilots a decisive edge when they came up against Egyptian, Syrian and Iraqi 'Fishbeds' in combat just a few months later in the Six Day War.

The enduring respect in which the Iraqi defector was held by Mossad was evident 32 years later, when veterans of the service who had been involved in Operation Diamond honoured Munir Redfa at a memorial service held after his death from a heart attack, aged 64, in August 1998.

Red Eagles

Besides Israel, the United States was also a beneficiary of Operation Diamond. With US airmen achieving disappointing results in air combat against North Vietnamese MiG-21s, the Americans were very keen to test the Soviet fighter in order to devise improved tactics. After high-level negotiations, the Israelis agreed to a loan of Captain Redfa's fighter. The American examination and evaluation of the aircraft was given the codename Project Have Doughnut (allegedly due to the MiG's doughnut-shaped nose air intake) and was conducted amid very tight security at Groom Lake in Nevada – popularly known as Area 51. The testing programme was conducted by the Foreign Technology Division, with a planned total of 102 sorties to be carried out by both USAF and US Navy pilots in the MiG-21 (redesignated YF-110 and painted with USAF markings) between 23 January and 8 April 1968. Of these, only eleven had to be cancelled due to maintenance issues, a serviceability level which 'US jets didn't come close' to, noted the evaluator.[13]

The American pilots and groundcrews praised the MiG's simplicity, reliability and ease of maintenance. In mock dogfights with the F-4

Phantom, its main antagonist over Vietnam, it was found that, while the F-4 could 'easily accelerate to above the usable airspeed of the FISHBED E' below altitudes of 15,000ft, the Soviet fighter had 'more instantaneous G available than the F-4 at any given airspeed.' The MiG-21 was also pitted against other US warplanes used in Vietnam, including the F-105 Thunderchief and F-100 Super Sabre. The F-105, warned the evaluators, was particularly vulnerable to the MiG, which enjoyed 'a distinct advantage in turn capability at all airspeeds and altitudes' over the Thunderchief and advised F-105 pilots to avoid prolonged manoeuvring engagements. Against the F-100, the Fishbed was superior in almost all respects.

On the negative side, the testers were critical of the instrument layout, which gave the cockpit 'a cluttered appearance' and could be 'confusing to an inexperienced pilot.' Additional shortcomings highlighted in their report were poor cockpit visibility, transonic vibration at low altitude, severe pipper jitter during cannon firing (which, said the test pilots, made it virtually impossible to track targets when pulling over 3Gs), a lack of armour protection around the engine, a radar that was highly vulnerable to ECM, and poor engine acceleration from idle to full military power.

Overall, however, the evaluators' verdict was that the MiG-21 was 'a good, honest aircraft' which 'represents a formidable threat to US tactical forces.'

Their opinion was that the Have Doughnut programme 'was very successful in obtaining a vast amount of invaluable data of an operational nature.' Among the test pilots' recommendations were that 'the Foreign Technology Division be authorized to aggressively pursue obtaining other foreign aircraft for exploitation.' They also suggested that figures from the US aircraft industry would benefit from participation in subsequent exploitation programmes. 'This,' their report stated, 'would allow our aircraft manufacturers the benefit of seeing the results of foreign technology first hand and give them a better understanding of the competition that we must meet and beat.'[14]

A few months later, the test pilots at the Foreign Technology Division were granted the opportunity to evaluate another mainstay of the VPAF, the MiG-17F 'Fresco', again courtesy of the Israelis. On 12 August 1968, two Syrian Air Force Lim-5s (the licence-built Polish copy of the MiG-17) landed by accident at a disused airstrip in northern Israel. The MiG-17s were interned and, after thoroughly testing the aircraft themselves, passed on by the Israelis to the Americans, whose technical exploitation of the type went by the codename Have Drill.

The first flight was made on 17 February 1969. The Americans found the Fresco to be 'extremely reliable', and in 224 flights carried out at Groom Lake with both aircraft 'only 23 major discrepancies occurred and only two missions were lost.' The test pilots noted that the aircraft possessed impressive manoeuvrability but that, due to its lack of hydraulically-powered controls, it became difficult to handle at speeds above Mach 0.85. 'Controls did not have pitch, roll, or yaw stability augmentation and only the ailerons were hydraulically boosted,' their report revealed.[15]

Knowledge gained on the capabilities and limitations of the MiG-17 and MiG-21 through Have Drill and Have Doughnut was incorporated into new training programmes for USAF and US Navy fighter pilots, such as the Navy's 'Top Gun' course, established in 1969 at the Fighter Weapons School at the Miramar naval air station in California. These courses were set up to improve the air combat skills of US airmen, whose shortcomings had been painfully exposed over North Vietnam.

The success of Have Drill and Have Doughnut eventually led to the creation by Air Tactical Command of a dedicated USAF unit to devise appropriate tactics to employ against Soviet fighters by pitting captured examples in mock aerial combat against American machines. Officially called the 4477th Test and Evaluation Flight and nicknamed the 'Red Eagles', it was formed under the codename Project Constant Peg – the name deriving from a combination of the callsign ('Constant') of General Hoyt Vandenburg Jnr, the USAF's Director of Operations and Readiness, and the name of the wife ('Peg') of the unit's first CO, Major Gaitlard Peck.

Activated on 1 April 1977, the 4477th was based at Nellis AFB and the Tonopah Test Range, both located deep in the Nevada desert, well away from prying eyes. The unit comprised an average of around 16 pilots, recruited mostly from the USAF's 'Aggressor' squadrons but also including some US Navy and Marine Corps flyers, and initially had a pair of MiG-17s and a single MiG-21 at its disposal.[16] This small fleet was soon bolstered by the addition of further MiG-21s bought from the Indonesian Air Force and other sources. The flying at Nellis and Tonopah was carried out under conditions of the strictest secrecy, with flights scheduled to avoid periods when Soviet satellites were known to be passing over the area.

But the MiG-17 and MiG-21 were by then obsolescent. The 4477th needed the latest Soviet hardware for testing, like the MiG-23 'Flogger'. As relations between Egypt and its traditional arms supplier, the USSR, soured in the 1970s, the CIA's station chief in Cairo, Jim Fees, saw an opportunity to meet the unit's needs. The Egyptian Air Force was a major operator of the Flogger, its first batch of MiG-23BNs becoming operational in 1974. Fees' initial requests to the Egyptians for a MiG-23 were rebuffed, but following the intervention of the country's vice-president Hosni Mubarak, permission was granted for a US test pilot to try out a MiG-23 in Egypt. Mubarak also allowed the CIA to copy the MiG-23's operator's manual, overruling the objections of the Egyptian military. Cooperation between Washington and Cairo deepened in September 1977 when Egyptian president Anwar al-Sadat approved the handover of a MiG-23BN to the Americans, which was disassembled and flown back to Nellis aboard a C-5 Galaxy. A further seventeen MiG-23s were later donated to the US, for use by the Red Eagles. The MiG-23BN was evaluated during 1978 under a programme codenamed Have Pad. The testing revealed that the Flogger was considerably faster than had been assumed, but suffered from poor manoeuvrability.

By the time Project Constant Peg was closed down in 1988, the 4477th had made over 15,000 flights and trained around 6,000 US airmen in simulated combat against its eventual collection of several dozen MiGs at Nellis and Tonopah. Although the 4477th was disbanded in 1990, secret testing of Russian aircraft is thought to continue at Nellis by a successor unit, Detachment 3 of the 53rd Test and Evaluation Group, with reported sightings of Sukhoi Su-27 Flankers flying from the base.[17]

Chapter Nine

Snoopers

nother means of gathering intelligence on Soviet bloc aircraft and aviation equipment during the Cold War was via the military liaison missions stationed in the GDR. Following the division of Germany after the Second World War into British, Soviet, American and French zones of occupation, on 16 September 1946 the UK and USSR signed a formal agreement authorizing each country to maintain a small military mission in each other's zone, the official purpose of which being 'to maintain Liaison between the Staff of the two Commanders-in-Chief and their Military Governments in the Zones' and 'make representations regarding their Nationals and interests in the Zones in which they were operating.' This was known as the Robertson–Malinin Agreement, after its signatories: Lieutenant General Brian Robertson and Colonel General Mikhail Malinin, the deputy military commanders-in-chief of their respective zones of occupation. The official title for the British mission was the rather long-winded 'British Commander-in-Chief's Mission to the Soviet Forces of Occupation in Germany', or British Exchange Mission, shortened to 'BRIXMIS' (and known to its personnel simply as 'the Mission'). Its Soviet counterpart was known as 'SOXMIS' (Soviet Exchange Mission).

The main provision of the Agreement was 'the freedom of travel and circulation for the members of Missions in each Zone with the exception of restricted areas, in which respect each Commander-in-Chief will notify the Mission and act on a reciprocal basis.'[1] As the Iron Curtain descended across Central Europe and the battle lines of the Cold War were drawn, both sides recognized that, with these liaison teams able to operate officially in what was now effectively enemy territory, the Agreement offered excellent opportunities for covert intelligence-gathering.

The original HQ of BRIXMIS was at Wilmersdorf in Berlin, but relocated in the late 1950s to the Olympic Stadium complex in Charlottenburg, while another BRIXMIS station, known as 'the Mission House', was set up in Potsdam in the Soviet zone. SOXMIS had its HQ at Bad Salzuflen before later moving to Bünde in Bad Oeynhausen. Each Mission was allocated 31 passes, for 11 officers and 20 other ranks. Initially, total BRIXMIS staff numbered around 80, eventually increasing to 200, comprising chiefly Army and RAF personnel, split on a ratio of roughly three to one, and commanded by a brigadier. The US and France joined the military liaison arrangement in April 1947 with the formation of the USMLM (United States Military Liaison Mission) and MMFL (*Mission Militaire Francaise de Liaison*). The American contingent was granted 14 passes and the French 18.

An air component was added to the BRIXMIS set up in 1956, when a de Havilland Canada DHC-1 Chipmunk of the RAF Gatow Station Flight was put at the unit's disposal. A two-seat basic trainer, the Chipmunk's official duties in Germany were liaison flying and to maintain the skills of RAF pilots, but BRIXMIS also used it to conduct spy flights at low-level, with an observer behind the pilot equipped with a 35mm Leica camera to photograph military installations along the designated air corridors that led in and out of East Germany. Occasionally, pilots would 'accidentally' stray outside these corridors so that their observer could photograph areas of interest. A second Chipmunk was added to the Station Flight in April 1968, to ensure that one would always be available for BRIXMIS tasking. Should a Chipmunk make a forced landing in GDR territory, the crew were under strict orders to destroy all maps and photographic equipment to conceal the true purpose of these flights. Suspecting that the

Chipmunks were carrying out aerial espionage, the Soviets and East Germans would often move sensitive equipment they didn't want photographed under cover when these aircraft approached.[2]

While its official function was military liaison, the real work of BRIXMIS and the other missions was espionage. BRIXMIS personnel carried out what were known as tours, travelling around the Soviet zone on foot and in marked cars, observing, photographing and, occasionally, even stealing the latest Warsaw Pact hardware at military bases throughout the GDR. Although these bases were designated PRAs (Permanently Restricted Areas) and officially off limits, BRIXMIS teams habitually breached this rule. The Army's contingent, naturally enough, would concentrate on obtaining intel on tanks, armoured vehicles, artillery and troop formations, while the RAF personnel focused on air bases, bombing ranges and aircraft. The Army and RAF teams operated separately until the 1970s, when joint operations were introduced to enhance cooperation.

The importance of the RAF's contribution was recognized in 1956 with the appointment of a Group Captain as Deputy Chief of Mission. The first to hold this post was Group Captain F.G. Foot, a Canadian Second World War veteran, who brought a much greater degree of professionalism to the unit's work, which hitherto had been, according to one BRIXMIS veteran, somewhat amateurish.[3] Out went hastily drawn sketches of target equipment, in came photography to confirm intelligence scoops.[4]

The personnel of the MLMs wore military fatigues and all were unarmed. The most important pieces of kit issued to the BRIXMIS teams were binoculars and Leica – later, Nikon – cameras, fitted with long-range lenses, to secretly photograph targets. The various electronic pods carried on the undersides of aircraft were of special interest to the watchers.

Spending days in a damp, cold OP observing airfields, hoping to snatch a glimpse of a new aircraft or item of hardware, was a monotonous and often thankless task. Nor was it without its dangers. The soldiers and airmen of BRIXMIS had to contend with the Stasi and their KGB and GRU allies, who monitored their movements closely. As many as 30 Stasi agents could be assigned to tail a BRIXMIS team, who would go to considerable lengths to shake off the 'narks', as the British called them. Sometimes elite Soviet *Spetsnaz* special forces teams were deployed to catch BRIXMIS members in the act of snooping within PRAs.

Encounters between the military liaison teams and the Soviets and East Germans were generally good-natured, their activities regarded as part of the 'game' of the Cold War. But occasionally encounters could take a violent turn, often at times of high international tension between the Soviet Union and the West. On 10 March 1962, for instance, a BRIXMIS car was fired on by East German border guards in Potsdam, close to the West German border, having apparently mistaken them for refugees trying to flee to the West. Several rounds hit the RAF driver, Corporal Douglas Day, who was badly injured but survived after having emergency surgery. Ten years later, on 6 September 1972, an RAF air tour spying on the Welzow air base was caught by Soviet 'narks', who knocked one of the team, Corporal Calvin, unconscious with the butt of an AK-47. In July 1987, Flight Lieutenant David Browne and Flight Sergeant Andy Peacock were assaulted and had their cameras confiscated by GRU officers while carrying out observation of helicopters equipped with ECM equipment. Deliberately forcing the vehicles of the MLMs off the road was another favoured tactic of the GDR security forces.

There were even occasions when helicopters were used to try to disrupt BRIXMIS surveillance ops. Flight Sergeant Colin Birnie was on a tour in the 1980s, photographing the aircraft weapons range at Retzow in Brandenburg, when he and his team were spotted by the crew of an Mi-24 'Hind' gunship, which flew towards them at ultra low-level. As Flight Sergeant Birnie and his team bundled into their

Opel Senator to escape, the Mi-24 bumped the car's roof with its undercarriage.[5]

The most serious incident occurred on 24 March 1985, when Major Arthur Nicholson, a member of the USMLM, was shot dead by a Soviet sentry after infiltrating a tank gunnery range at Ludwigslust to take photographs of the Soviets' new T-80 Main Battle Tank.

Attacks and harassment by the Soviets were not confined to ground tours. 'Buzzing' of the light aircraft operated by the MLMs while flying in the air corridors by VVS and LSK fighters and helicopters was relatively common. 'The MiGs used to routinely buzz us to mess with us,' revealed US Army Captain Tim Bloechl, who flew corridor flights in a Pilatus UV-20A in the mid-1980s. 'We sat sideways in the plane and opening the door – left to right – to hang out and take pictures.'[6] Trigger-happy Soviet and East German soldiers also sometimes took potshots with their AK-47s at aircraft of the MLMs flying over their bases, and there were incidents of Chipmunks arriving back at RAF Gatow with ballistic damage.[7]

But the personnel of the MLMs were often rewarded for their troubles with some notable intelligence scoops. One of the first of these was in the summer of 1957. Group Captain Foot, the Deputy Chief of BRIXMIS, and his assistant Squadron Leader Neubroch were observing the airfield at Welzow when they spotted one of the newly introduced Yak-25 'Flashlight' interceptors with its nosecone removed, revealing its highly secret RP-6 Sokol radar. According to Neubroch, the photographs they took of the radar gave NATO intelligence analysts valuable information on the air intercept capabilities of the Yak-25.[8]

During 1974, BRIXMIS took the first photographs of the MiG-23 'Flogger', Mil Mi-24D 'Hind' and Tupolev Tu-22 'Backfire' strategic bomber in East Germany. In April 1978, pictures were taken of the Tupolev Tu-143 unmanned reconnaissance drone, which had entered service two years earlier, and a few months later obtained photographs of the AT-2 'Swatter' ATGM (Anti-Tank Guided Missile), carried on the stub wings of a 'Hind' gunship.

When NATO received reports in 1980 indicating that the Soviet Air Force planned to deploy the latest variant of the 'Flogger', the 'G' model, to the GDR, all three Allied MLMs joined forces for a major combined surveillance operation to obtain photographs of the aircraft. After three months, the watchers' patience finally paid off when a BRIXMIS member snatched a photograph of the MiG-23ML 'Flogger-G' and its reconnaissance pod at an East German air base. A BRIXMIS team also took the first images of the Soviets' most advanced fighter, the MiG-29 'Fulcrum', at the Wittstock air base in 1986, and two years later secured photographs of the latest reconnaissance pod carried by a MiG-23 'Flogger-J'.

The watchers were also able to dispel certain myths regarding the proficiency of Soviet pilots and the capabilities of their machines through observation of training exercises at air bases, including the widely held belief in NATO circles that the 'Hind' gunship could not operate effectively in wintry conditions.

The Chipmunk crews of the BRIXMIS Flight enjoyed a number of intelligence scoops, too. One of the most significant of these was in the early 1970s, during a routine sortie to photograph the airfield at Zerbst, 100 miles southwest of Berlin, along the southern corridor. Expecting to see the usual line-up of MiG-21s stationed at the base, the observer was surprised when he spotted what looked very much like a McDonnell Douglas F-4 Phantom sitting outside a hangar. Analysis of the photographs he took by RAF Photo Interpreters confirmed that the aircraft had no national markings and that the nose section looked non-standard. In their book *Looking Down the Corridors*, Cold War historians Dr Kevin Wright and Peter Jeffries stated:

Much speculation followed as to how and why it had got there. The assessment was that it was a former USAF aircraft that had been shot down in Vietnam and that the Soviets had made it sufficiently airworthy to take round to brief their own fighter pilots on its strengths and weaknesses.[9]

Besides photography and visual reconnaissance, there were rare occasions when the opportunity presented itself to MLMs to infiltrate airfields and scoop up items of military hardware right from under the noses of the Soviets and East Germans.

In 1970, a BRIXMIS tour led by Squadron Leader Rod Saar, covertly filming MiG-21S 'Fishbed-Js' carrying out gunnery and bombing practice at a range in Belgern with a cine camera, sneaked onto the range and collected spent cannon shells from the MiGs for analysis by British weapons specialists. Saar and his team followed this with an even more audacious snatch mission – stealing a UXB (Unexploded Bomb) from the range. After carrying it between them to their car, the team found that the bomb (which was almost five feet in length) wouldn't fit in the boot and so drove it back to BRIXMIS HQ at Charlottenburg lying on the rear passenger seats. After taking the bomb into an office in the HQ building, the UXB was made safe by an EOD expert. Saar was advised by the BRIXMIS CO to refrain in future from bringing unexploded ordnance into the office.[10]

USMLM teams could be just as adventurous as their British counterparts in pursuit of discarded aviation hardware from military bases. From 1988, the LSK began receiving MiG-29 'Fulcrum' fighters from the Soviets. Soon after, one of these crashed on landing at an unnamed East German air base. A USMLM team spent several days in a concealed OP near the base, covertly observing Soviet and East German personnel collect pieces of the wreckage and burying the remains of the MiG-29 near the airfield. Once the coast was clear, the American watchers reportedly breached the airfield's defences, dug up the buried parts and smuggled out some of the wreckage. Samples of the aluminium alloys and composite materials used in the Fulcrum's fuselage proved to be of particular interest to the US technical experts who later studied them. 'Analysis of the material indicated the Soviets were ten years further advanced in their metallurgy progress than the United States had believed,' revealed the military historian Colonel Richard Camp USMC.[11]

The 'Firebar' coup

What is generally regarded as the greatest intelligence coup carried out by an MLM took place in April 1966. On the afternoon of the 6th, a pair of Yak-28P interceptors of the 24th Air Army's 35th IAP were flying from Finow to Köthen air base when one of the Yaks suffered catastrophic engine failure while flying over Berlin. Realizing he would not be able to make it to an airfield in the Soviet zone, the pilot Captain Boris Kapustin requested permission from the Soviet air controller to attempt an emergency landing at a NATO air base. This was denied and, at 2.30pm, the Yak-28 crashed into the Havel River in the British sector of West Berlin. Neither Kapustin nor his navigator, Lieutenant Yuri Yanov, survived.

After receiving reports of the crash, the West Berlin police and soldiers of the Royal Military Police rushed to the scene to cordon off the area around the crash-site. When BRIXMIS was alerted, they, too, sent a team to the site. Soon, divers from the West German fire department arrived to begin a search for the crew, while a British military helicopter hovered overhead. Within hours, a two dozen-strong detachment of armed Soviet soldiers from the ceremonial guard at the war memorial at Tiergarten was despatched to the Havel, and a standoff ensued.

As the situation at the crash-site grew more tense, a high-level meeting of senior British military commanders, among them the C-in-C of BRIXMIS Brigadier David Wilson, was held at the British HQ in Berlin to decide their next steps. During this meeting a Soviet delegation headed by a colonel arrived at the HQ building and was told to immediately withdraw their armed troops from the scene of the crash. They reluctantly agreed, but insisted that a party of unarmed soldiers be permitted to remain as observers. The British commanders also agreed at the meeting to issue a formal protest to the Soviets deploring 'this latest example of irresponsible and careless flying over heavily populated areas which endangered the lives of the citizens of Berlin.'[12]

Behind the confected diplomatic outrage, the British could not believe their good fortune, for they now had a golden opportunity to examine what the CIA called in a report from the time as the 'most advanced fighter in the Soviet operational inventory.'[13]

The Yakovlev Yak-28 project was instigated in the mid-1950s to meet a requirement for a supersonic tactical strike aircraft, which was developed under the designation Yak-129. Powered by two Tumansky R-11 afterburning turbojet engines slung beneath its swept wings, the Yak-28 made its maiden flight on 5 March 1958 and entered Soviet service two years later. NATO observers gained their first glimpse of the Yak-28 at the Tushino air display in 1961, where they initially believed it to be a modified variant of the Yak-25 'Flashlight' interceptor. After realizing that it was in fact an all-new aircraft intended for the strike mission, they assigned it the reporting name 'Brewer'. Reconnaissance and electronic warfare variants were subsequently developed, and were known to NATO as the 'Brewer-D' and 'Brewer-E', respectively.

The versatile Yak-28 airframe was also adapted into a long-range, all-weather air defence interceptor, fitted with the powerful RP-11 Oryol-D air intercept radar, which had a maximum range of 30km. The interceptor variant emerged in 1964 as the Yak-28P, which was christened 'Firebar' by NATO. The 'Firebar' was also powered by two R-11 engines, giving it a maximum speed of 1,143mph and a ceiling of 55,000ft. Range was 1,550 miles. In keeping with prevailing air combat doctrine at that time, the 'Firebar' eschewed cannon in favour of all-missile armament – usually two long-range R-98M (AA-3 'Anab') radar-guided and two short-range K-13 (AA-2 'Atoll') heatseeking missiles. A USMLM team first identified the Yak-28P in East Germany in March 1966, at the Finow air base, 30 miles from Berlin.

After the search for the bodies of the two crewmen in the Havel was called off for the night, RAF operations officer Squadron Leader Maurice Taylor took photographs of the tail of the aircraft, which was protruding from the surface of the water. From a close study of these, RAF Junior Technician Eddie Batchelor was able to positively identify the aircraft from two distinctive aerials attached to the tail as a Yak-28P. A top priority 'flash' signal was then sent to the Defence Intelligence Secretariat in London and it was decided to make an all-out effort to recover and secretly examine as much of the wreckage of the ill-fated interceptor as possible.

This would present the British with both a major logistical and diplomatic challenge, and would have to be carried out as discreetly as possible, under the watchful gaze of the suspicious Soviet observers gathered on the banks of the river. At first light the next day, a heavy ferry equipped with a crane of the Royal Engineers' 38 Field Squadron sailed up the Havel and anchored next to the crash-site. Divers led by Captain Roger Eagle of the Royal Engineers made repeated dives on the wreck, joined by Royal Navy divers flown out from Portsmouth. They found the bodies of Captain Kapustin and Lieutenant Yanov still in their seats, having failed to eject. Once the bodies had been removed, the divers used cutting equipment to remove the Oryol-D radar and other items of interest from the fuselage. They were then brought to the surface and loaded on the ferry. Technical experts were flown in from RAE Farnborough to examine the salvaged parts.

While the carcass of one of their most advanced warplanes was being systematically stripped by the British, the Soviets – including Major General Vladimir Bulanov of the Soviet Air Force and Colonel Lezzov, the deputy C-in-C of SOXMIS – could only look on with mounting frustration and issue dark warnings that they would be 'forced to take appropriate measures' if the aircraft wreckage and the bodies of the crew were not swiftly handed over.[14] The Americans were also taking a close interest in the British salvage operation, the CIA reporting to the US President on 7 April that 'major portions of the aircraft are intact and should tell us much about the equipment aboard.'[15]

Colonel General Pyotr Belik, deputy commander of Soviet Forces in Germany, delivered an angry protest in person to Brigadier Wilson, accusing the British of 'deliberately delaying the salvage operations', of being 'very slow in bringing out the bodies' and of 'removing pieces of the aircraft and "dividing up the aircraft under the water to examine details."' He also insisted that 'if the Russians had been allowed to carry out the salvaging of the aircraft they would have done it much quicker' and complained that they 'would never be able to find out what had gone wrong with their aircraft because it had now been so badly mangled.' He rounded off his protest with a demand that the British immediately hand over the aircraft. Brigadier Wilson mollified General Belik somewhat by assuring him that he hoped to inform him later that day when the Yak-28 could be returned.

Meanwhile, the British kept playing for time. 'We have devised a salvage deception plan,' BRIXMIS informed the UK embassy in Bonn, 'which may be of some slight value in keeping the Russians guessing about whether salvage operations are still going on.'[16] The Soviets also played their own tricks to disrupt the recovery effort, including shining powerful searchlights from their trucks parked on the shore at the salvage party at night, forcing them to stop working for a time.

In the early hours of Good Friday, 8 April, the bodies of Captain Kapustin and Lieutenant Yanov were handed over to the Soviets (though not before all the documents found in their flying suits had been carefully photographed) in a solemn ceremony, attended by a party of senior Soviet and British military officers, with a piper of the Royal Inniskilling Fusiliers playing the lament. The recovery and handover of the bodies did much to reduce the tensions that had built up, General Belik even complimenting the bearing of the British troops who provided the guard of honour.

Believing they could no longer plausibly delay the handover of those parts of the aircraft that had been successfully recovered, Brigadier Wilson informed the Soviets that they would be returned the next day, at 1800 hours.

The most important components yet to be located by the British salvagers were the two Tumansky engines, which had broken off the wings upon hitting the water. After much searching in the dark, a British diver finally found the engines buried in the mud on the riverbed. Strops were tied around the engines and they were secretly dragged just below the waterline to a jetty a mile away. Once safely out of sight of the Soviet observers, the engines were raised and taken to RAF Gatow for an initial inspection by British and American engineers. They were then flown to Farnborough for more detailed study. The tips of the engines' turbine blades were also sawn off so that their metallurgical composition could be determined. Forty-eight hours later, the engines were flown back to Germany and just as covertly returned to the Havel, where they were 'discovered' soon after.

A few days later than promised, the wreckage of the Yak-28P was handed over to the Soviets on a raft. Inspecting the parts, General Bulanov noted the shorn off ends of the engines' turbine blades and that the dish of the Oryol-D radar unit – which the British had decided to hang on to – was missing. When quizzed about the latter, the British feigned ignorance and insisted that all parts successfully recovered were present and correct. Though clearly unconvinced, Bulanov decided not to press the matter.

The salvage and secret examination of the 'Firebar' was perhaps the greatest intelligence success involving BRIXMIS during its 44-year existence. The organization was finally disbanded on 31 October 1990, as the Cold War came to a close.

Although much useful intel was gleaned from the wreckage, particularly about Soviet A.I. radar, the Yak-28 itself had a fairly undistinguished service career. Never regarded as an entirely successful design by the Soviets, total production (all variants) amounted to 1,180 – modest by Soviet standards – and it served with only a handful of PVO units, based in the far north of the USSR, and two squadrons of the 24th Air Army, forward deployed in East Germany. It was phased out of frontline service in the early 1980s, with the final examples retired shortly after the dissolution of the Soviet Union.

Chapter Ten

Hot War

Throughout the Cold War, the threat of Mutually Assured Destruction – or MAD, as it was commonly and aptly known – ensured that, with their vast stockpiles of nuclear weapons, there would be no direct military confrontation between NATO and the Warsaw Pact – although on occasion the world came perilously close to nuclear war, such as during the Cuban Missile Crisis of 1962. Instead, the Cold War years were punctuated with many regional 'hot' wars, limited in scope and confined to their locality, fought out mainly in the Third World, with East and West backing the opposing forces based on their ideological orientation, including with the supply of their latest military hardware. These conflicts offered both sides excellent opportunities to lay their hands on their opponents' warplanes, either those shot down or captured during or after hostilities.

The first of these Cold War hot wars was Korea. As we have seen, the MiG-15's introduction into the Korean War prompted a huge effort by both the British and Americans to acquire one of these jet fighters from North Korea (see Chapter Five). The Soviets, of course, were just as keen to obtain examples of the aircraft the US deployed to the Korean peninsula.

The Korean War saw the first widespread use in combat of the helicopter, which proved to be especially useful in the search and rescue and casualty evacuation roles. Rotary-wing aircraft was an area of aviation where the Americans enjoyed a significant lead on the Soviets. Dissatisfied with the state of helicopter development in the Soviet Union, in 1951 Stalin summoned the heads of the country's leading OKBs charged with producing rotary-wing aircraft – Aleksandr Yakovlev, Nikolai Kamov and Mikhail Mil – to the Kremlin to make his displeasure known and demand that progress be speeded up. To achieve this, Soviet military advisers serving in North Korea conceived a plan to lure a US helicopter crew into the North and capture their aircraft. They coerced a USAF fighter pilot who had been shot down and captured in North Korea to radio a distress call, requesting a helicopter pick-up. A Sikorsky H-19 Chicksaw, the most modern helicopter in the American inventory, was quickly despatched to the downed pilot's supposed location. Soviet military liaison officer Colonel Valentin Sozinov explained what happened next:

> There was a little clearing and the helicopter started to land. Right then something happened. It looked as if the pilot had only just landed and started to climb very fast. At that point we opened fire. The fire was hot. The rotors were damaged, and the helicopter fell from about 50 or 70m. It fell on its side when it hit the ground.[1]

Fortunately, the H-19 suffered only relatively minor damage in the crash and the crew survived, to be taken prisoner. The Soviets and their North Korean allies dragged the helicopter away into cover. A Soviet aide to North Korea's leader Kim Il Sung, Valentin Pak, added that Moscow ordered the H-19 be dismantled and shipped back to the TsAGI in Moscow for detailed analysis. 'I heard personally how they took it apart and packed it into crates, and everything literally down to the last screw was sent immediately back to Moscow,' he recalled. Pak also revealed that the military officer in charge of the operation was awarded the Order of the Red Banner for this success.[2]

The greatest prize for the Soviets was, of course, the North American F-86 Sabre, the MiG-15's main antagonist. Some useful information about the American jet fighter was gleaned through interrogations of Sabre pilots shot down over North Korea and taken prisoner. In total, some 262 captured USAF airmen were questioned by Soviet and Chinese intelligence officers during the course of the war. 'Essentially the questions were about tactics,' revealed one Soviet officer who served in Korea. However, information on highly secret equipment fitted to the F-86, including its state-of-the-art AN/APG-30 radar gunsight, was also extracted from PoWs. So important was this information considered to be that some reports on USAF prisoner interrogations were said to have been read by Stalin himself.[3]

Useful as PoW interrogations were, what the Soviets really coveted was an intact, flyable example of the F-86, with Stalin reportedly issuing the order personally for one to be captured and brought back to Moscow. This task was initially assigned to a specialist MiG-15 unit made up of a dozen test pilots from the NII at Moscow, under the command of Lieutenant Colonel Dzyubenko, which arrived in North Korea in April 1951 and was attached to the 196th Fighter Air Regiment, based at Andung. Dzyubenko's plan was for his team to use their MiG-15s to box in an F-86 and force its pilot to land on North Korean territory, and the pilots spent a month practising their formation flying. But the first attempt to put this plan into action backfired. On 31 May, Dzyubenko's pilots attacked four F-86A-5s escorting two B-29 bombers over Anju in South Pyongan province. The Americans mounted a spirited counter-attack, claiming three MiGs shot down with no loss to themselves in the ensuing air battle – two by F-86s and one by a B-29 gunner (Soviet records insist only one MiG was lost in the engagement). Further misfortune befell the squadron a few days later when Dzyubenko was killed in a landing accident. The unit was disbanded soon afterwards.

The Soviets finally got lucky on 6 October 1951, when Second Lieutenant Bill Garrett made an emergency landing in an estuary on the North Korean coast, after his F-86A-5 was damaged in a skirmish with a MiG-15 flown by Lieutenant Colonel Yevgeny Pepelyaev, one of the top-scoring Soviet aces of the Korean War. Lieutenant Garrett was swiftly picked up by a Grumman SA-16 amphibian. Just as the British and Americans had done a few months earlier in Operation MIG (see Chapter Five), a major salvage operation was quickly launched by the Soviets and Chinese to recover the downed jet. During this salvage operation, they were spotted and fired on by US warships, but the Soviet-Chinese team succeeded in recovering the wreckage. The pieces were then loaded into trucks and driven back to Andung air base. This journey was not without incident, one of the trucks being strafed by a roving USAF B-26 and forced to take cover in a tunnel.

Before being sent on to Moscow, at Andung Colonel Pepelyaev took the opportunity to sit in the F-86 and was impressed by the layout and comfort of the cockpit, comparing it to a luxury car.[4]

At the LII, the F-86A was dissected and intensively analysed, with each component meticulously measured and photographed. Serious consideration was given to reverse-engineering the F-86 and producing an exact duplicate. But this plan was eventually abandoned in favour of incorporating the Sabre's various technical innovations into the MiG-15's successor, the MiG-17. Another intelligence windfall came the Soviets way in July 1952, when Colonel Walker 'Bud' Mahurin, a Second World War ace and CO of the 4th Fighter-Interceptor Wing, made a forced landing in a rice paddy after his F-86E was hit by AA fire during a strafing run. The virtually intact fighter was swiftly recovered and also brought back to the LII.[5]

Entering USAF service in early 1951, the 'E' model was the latest variant of the Sabre. Among the technical secrets this Sabre revealed to the Soviet analysts were fully powered flight controls and a hydraulically actuated 'all-flying' tailplane, which greatly enhanced manoeuvrability. This system would first be adopted by the Soviets in their MiG-19 fighter, which entered frontline service in 1955. For the Soviets, one of the most important items of technology found on the F-86E was the Sperry AN/APG-30, the radar-directed gunsight they'd first learned about through PoW interrogations, which could accurately track a target up to a range of almost 2,800 metres. The Soviets wasted no time in copying the device and installing it in their own fighters, as the ASP-4N.

The Soviets also based their PPK-1 g-suit on the G-3A model worn by F-86 pilots captured over North Korea.[6] By preventing black out during high-g manoeuvres by maintaining the flow of blood to the brain, the g-suit gave American fighter pilots a crucial advantage over their communist adversaries in dogfights during the Korean War.

The missile age

The increasing speeds of jet aircraft in the 1950s led to a belief among the air chiefs of the major powers that shooting down enemy fighters with machine guns and cannon would become all but impossible. If air combat was to remain viable, a new weapon would have to be devised.

Inspired by Nazi research captured at the end of the Second World War, in the late 1940s a team of engineers under the leadership of physicist William B. McLean began development of an air-to-air missile at the Naval Ordnance Test Station at China Lake in the Mojave Desert. Aptly named the Sidewinder, after a breed of rattlesnake native to the area that hunts prey through its heat, it was an infrared-homing missile with a range of almost three miles and a warhead containing 4.5kg of explosives, with a blast radius of around nine metres.

Although the missile could only engage targets from the rear and was susceptible to locking on to alternative heat sources, such as the sun, test firings against target drones carried out in 1953 were sufficiently encouraging for the US Navy top brass to order the AIM-9B Sidewinder into production. In May 1956, the missile began entering service with the US Navy.

It wasn't long before the Sidewinder was tested in actual combat. Rising tensions between communist China and the neighbouring

US-backed island of Taiwan (also known as the Republic of China) erupted into aerial battles between their air forces over the Taiwan Strait in August 1958. On paper, the Chinese PLAAF held a clear advantage, their Shenyang J-5s (the Chinese-built version of the MiG-17) boasting superior performance to the F-86F Sabres of the ROCAF. To level the playing field, in a secret operation codenamed 'Black Magic' the US Navy rushed a consignment of AIM-9B Sidewinders to the Taiwanese.

The new missile soon made an impact in the air war. On 24 September 1958, 16 PLAAF J-5s encountered a dozen ROCAF Sabres armed with Sidewinders over the Taiwan Strait. In the ensuing dogfight, the Taiwanese claimed nine J-5s shot down – four by Sidewinders – for only a single loss to themselves. Four days after the Sidewinder's successful debut, in another clash between the Chinese and Taiwanese air forces, the missile again proved its effectiveness when Sidewinder-armed F-86Fs shot down several more J-5s. This time, however, one of the Sidewinders failed to explode after becoming lodged in the fuselage of a J-5. The damaged J-5 managed to regain its base on the Chinese mainland and the unexploded missile was carefully removed from the aircraft.

The Chinese initially considered reverse-engineering the Sidewinder themselves, but after realizing this was beyond their technical capabilities invited their Soviet allies to take on the task, in return for a promise that all technical information gleaned from the weapon would be shared with them. As their own air-to-air missile, the K-5 (known to NATO as the AA-1 'Alkali') – which utilized a radar 'beam-riding' guidance system – had much less combat capability than the Sidewinder, being unable to track manoeuvring targets, this offer was gratefully accepted by the Soviets.

Taken to a missile research facility near Moscow, the Sidewinder was taken apart by a team of Soviet engineers under Ivan Toropov and its secrets quickly uncovered. The intelligence value of the captured

AIM-9B to their own guided missile programme was revealed by one of the senior engineers involved in analysing the Sidewinder, Gennadiy Sokolovskiy, who later described it as 'a university offering a course in missile construction technology which has upgraded our engineering education and updated our approach to production of future missiles.'[7]

Within a few months, Sokolovskiy and his colleagues had produced their own version of the Sidewinder, the Vympel K-13, a near-identical copy. Given the reporting name AA-2 'Atoll' by NATO, the K-13 was first test-launched from a MiG-19 in 1959 and began equipping VVS fighters from 1961. The 'Atoll' became the standard short-range AAM for the Warsaw Pact air forces for the next twenty years. In 1966, the Chinese began production of their own version, the PL-2. A CIA intelligence report from October 1965 stated that the Atoll 'is a relatively simple missile and could be installed on almost any jet fighter capable of achieving the Mach 0.8 speed necessary for launch.' Although 'accurate and reliable', like the Sidewinder it was based on, the Atoll's effectiveness 'drops sharply against bright backgrounds, in clouds or haze, or when the target is near the direction of the sun.'[8]

Treasure in the jungle

The largest and bloodiest of the many proxy wars fought out between the superpowers during the Cold War was in a small former French colony in south-east Asia: Vietnam. Following the departure of the French in 1954 after a bitter eight-year struggle against the ultimately victorious native Vietminh guerrillas, the country was divided in two: communist-ruled North Vietnam, backed by the USSR and China, and the nominally democratic South Vietnam, backed by the US. No sooner had the French withdrawn, however, than a civil war engulfed the country, as the North set out to unite the two Vietnams under their communist rule. Washington, fearing communist expansion in the region, supported the embattled South Vietnamese regime, initially by way of advisers and military aid, but escalating into full-scale, direct military involvement, with over half a million American troops committed to the fighting in the country's dense jungles by 1968.

With the US also deploying thousands of its latest strike aircraft for a sustained bombing campaign over North Vietnam (Operation Rolling Thunder), many of which fell victim to sophisticated anti-aircraft defences supplied by Moscow, like the formidable S-75 *Dvina* surface-to-air missile system (known to NATO as the SA-2 'Guideline'), the war offered plentiful opportunities for the Soviets and Chinese to acquire US aviation technology from shot-down aircraft. 'Since during the Vietnam War the Americans lost over 4,000 aircraft, there was a lot of work to do and we were never idle,' said Aleksandr Anosov, a member of a group of Soviet specialists, nicknamed the 'Wild Division', who were tasked with salvaging important components from the wreckage of downed US aircraft and shipping them back to the USSR for further study.[9]

Based at the Soviet embassy in Hanoi, the unit was hindered in this task by Vietnamese villagers, who would often strip a downed aircraft of anything of value before the technicians reached the crash-site. Besides the locals, the Soviets also had to contend with their Chinese counterparts, with an intense rivalry developing between the two communist powers to recover components from aircraft wreckage.

Flagrant breaches by the Chinese of an agreement with Hanoi that captured equipment should first be handed over to the local NVA commander strained Sino-Vietnamese relations, however. In 1966, having failed to first consult their hosts, the North Vietnamese stopped the Chinese from shipping a US reconnaissance aircraft back to China. On another occasion, Chinese troops of the 1st Railroad Engineering Division found a shot-down Ryan AQM-34 reconnaissance drone and, ignoring demands from NVA commanders that they hand it over, sent it under armed guard in a truck convoy to the Chinese-Vietnamese border. When this news reached General Vo Nguyen Giap, C-in-C of the North Vietnamese Army, he exploded: 'This is Vietnam! [We] must get it back!' The convoy was halted at the border by NVA troops and the Chinese were relieved of the drone.[10]

Despite this mutual rivalry among the communist allies, both the Chinese and the Soviets enjoyed considerable success in their technology acquisition efforts. In November 1967, a Republic F-105D Thunderchief that was brought down by an S-75 missile was sent back to the Soviet Union. Though not conclusive, evidence suggests the Soviets also acquired in Vietnam a near-intact McDonnell Douglas F-4 Phantom, the standard fighter of the USAF, US Navy and Marine Corps during the war, as well as equipping the air forces of numerous NATO air forces. The Chinese, meanwhile, 'took back control panels, communication devices, radar equipment, electronic systems, weapon controls, missiles, launchers, and cameras from crashed airplanes. Sometimes they dismantled the entire warplane and took the parts back to China,' according to historian Xiaobing Li.[11]

Perhaps the most important equipment recovered from downed aircraft related to the vital area of ECM. When the US commenced Operation Rolling Thunder in 1965, losses to North Vietnamese radar-directed AAA guns and S-75 missiles were unexpectedly heavy, particularly amongst the F-105 force. In an effort to reduce losses, the Americans turned increasingly to electronic counter-measures – equipment carried on aircraft which was designed to jam enemy radars and communications that controlled the SAM and AAA batteries.

The main ECM deployed in the early years of the air war over North Vietnam was the QRC-160 jamming pod. Just over eight feet long and ten inches wide, it was carried on an aircraft's underwing pylon and emitted signals to jam enemy fire-control radars. Following some initial trials in Vietnam in 1965, in September 1966 F-105Ds of the 355th Tactical Fighter Wing began routinely operating with the pods. Results were encouraging and led to a wide rollout of the QRC-160. By December 1966, all F-105s' flying missions over targets defended by S-75s carried the pods. The effect was dramatic, losses being slashed by around two thirds. The commander of the 388th Tactical Fighter Wing, Colonel William Chairsell, said that the QRC-160 was 'one of the most effective operational innovations I have ever encountered. Seldom has a technological advance of this nature so degraded an enemy's defensive posture.'[12]

For the defenders, unlocking the secrets of the enemy's anti-radar technology now became a priority, in order to develop their own electronic counter-counter-measures. Much was learned about the Americans' ECM capabilities from the wreckage of a Douglas EB-66C of the 41st Tactical Electronic Warfare Squadron, which was shot down by an S-75 over Bac Thai Province on 4 February 1967, killing three of its six-man crew. The EB-66C was a dedicated ECM and Elint platform, bristling with advanced anti-radar and communications equipment. According to the official North Vietnamese history of the war, 'The wreckage of this EB-66C yielded numerous documents and provided extensive information on enemy electronic warfare.'[13] The unit responsible for bringing down the EB-66C, the 89th Battalion of the 274th Missile Regiment, was awarded North Vietnam's highest military decoration in recognition of this success.

What was really needed, however, was an intact QRC-160 pod. A year later, on 14 February 1968, an F-105D was brought down near Hanoi by shrapnel from an S-75 missile, killing its pilot. Amongst the wreckage, the North Vietnamese found an almost undamaged QRC-160 pod. 'The secrets of the enemy's jamming of our missile guidance channel, which had caused us so much heartache, now lay revealed right there in front of our eyes,' remarked one senior North Vietnamese officer.[14] Passed on to the Soviets, this prize allowed their engineers to modify the S-75's 'Fan Song' radar to at least partially nullify the effectiveness of the jammer.

The technological advantage swung back and forth between attacker and defender for the next five years – the US introducing improved ECM equipment and tactics, the communists countering with improvements to their anti-aircraft systems – until the US ended offensive air operations over North Vietnam in January 1973.

In 1968, the Soviets also retrieved an AIM-7 Sparrow air-to-air missile from a US fighter that came down on North Vietnamese territory. Unlike the heatseeking Sidewinder, the Raytheon Sparrow used radar guidance to home in on its target and was designed to take out enemy aircraft at much greater ranges – up to 16 miles. A Soviet copy, designated the K-25, was developed by engineers at the Vympel OKB, though the project was cancelled in 1971 as another AAM then in development, the R-23, was judged to possess superior performance and be more resistant to anti-missile countermeasures. Technology derived from the captured Sparrow did, however, find its way into the design of the later Vympel R-27 AAM, which began arming Soviet fighters from 1983.

As in Korea, PoW interrogations could also pay dividends. From one such interrogation of a captured US airman, the communists learned some of the secrets of the Americans' AGM-62 Walleye. One of the first 'smart bombs', the Walleye was a television-guided glide bomb, with a 375kg warhead. One of the first missions in which Walleye bombs were used was an attack by US Navy A-4 Skyhawks on Hanoi's main power plant on 19 May 1967. In a follow-up strike three weeks later, one of the Walleyes dropped on the power plant failed to detonate and was taken away by North Vietnamese engineers, to be studied by the Soviets and Chinese.

Vietnam, of course, became a quagmire for the US. Mounting casualties, few clear battlefield successes, and a growing anti-war movement back home eventually led to American disengagement and thereafter the collapse of the beleaguered South Vietnamese regime. In April 1975 communist forces overran South Vietnam, taking the capital Saigon and ending the long, bloody war which had cost an estimated two million lives. On South Vietnamese Air Force air bases, the North Vietnamese found many intact aircraft supplied by the Americans as part of Washington's policy of 'Vietnamization', and some of these were passed on to the Soviets. Among them were examples of the ubiquitous Bell UH-1 'Huey', the Boeing CH-47 Chinook and the Cessna A-37 Dragonfly light attack jet.

The most advanced type operated by the South Vietnamese Air Force was the Northrop F-5 Freedom Fighter. A twin-engined, lightweight supersonic fighter, the F-5 was conceived as a relatively low-cost alternative to the more expensive and mechanically complex US fighters of the era, like the F-4 Phantom, to equip the air forces of US allies with modest defence budgets. The South Vietnamese received an initial batch of 19 F-5A and B models in June 1967, equipping the 552nd Fighter Squadron, this force growing to over 100 by the early 1970s as the process of 'Vietnamization' gathered pace. The NVA captured a number of the improved F-5E variant at the Bien Hoa air base when it fell in April 1975 and at least one of these (serial number 73-00807), along with spare engines and parts, was shipped back to the USSR.

The F-5E was sent to the aeronautical research and test centre at the Akhtubinsk air base, where it underwent a thorough evaluation. The Soviet engineers were impressed by the design, and particularly appreciated its ease of maintenance. A comprehensive programme of flight tests was then carried out from 1976-77, which included putting the F-5E up against the nearest Soviet equivalent, the MiG-21, in mock dogfights. The results of these revealed the American machine to be superior, Soviet test pilot Vladimir Kondaurov reporting that the MiG lost every engagement to the more manoeuvrable F-5E. Design features of the F-5, such as its instrument panel, were said to have influenced the development of Sukhoi's Su-25 'Frogfoot' ground-attack aircraft and Su-27 'Flanker' fighter.

Project MEXPO

Israel's decisive military victory over Egypt and its allies in the Middle East in the Six Day War (5 – 10 June 1967) left the Jewish state in possession of vast quantities of captured military hardware, mostly of Soviet origin. Included amongst this haul was the wreckage of shot-down aircraft, air-to-air missiles, surface-to-air missile batteries, radars and even intact fighters. The Israeli government appreciated the political as well as the practical value of the materiel that had fallen into their hands, much of which would be of great interest to the United States. Strengthening ties with the US was a key foreign policy goal of Tel Aviv at the time. Shortly before the Six Day War, French President Charles de Gaulle had imposed an arms embargo on Israel. This had particularly serious implications for the Israeli Air Force, which was mostly equipped with French aircraft, like the Dassault Mirage III and Sud Aviation Vautour bomber. It was therefore imperative that Tel Aviv forge a closer relationship with Washington to secure a new source of arms.

After negotiations between US and Israeli officials, it was agreed that several American technical teams from both the CIA and DIA would be despatched to Israel to conduct a joint examination of the captured equipment, the first of these teams arriving in early August 1967. The CIA's mission was known as Project MEXPO, while the DIA's contribution was codenamed Project Switch Hit.

There would be three phases to the two projects: Phase A involved a joint US-Israeli exploitation of the hardware on-site in Israel; Phase B was the export of certain pieces of equipment to the US for further study; finally, Phase C was to be the 'gift or sale to the US of selected items of equipment for US retention.' However, on 5 September Lieutenant General Joseph Carroll, director of the DIA, submitted a status report to the US Deputy Secretary of Defense Paul Nitze expressing frustration that while the Israelis had given the American teams access to 'low priority items,' such as land mines and communications equipment, they had 'demonstrated a marked hesitancy in allowing the technical specialists access to higher priority items.' Amongst these were the S-75 *Dvina* SAM system and its 'Fan Song' radar. As the S-75 was responsible for almost a third of all US air losses over North Vietnam, examination of a captured example was accorded the highest priority by the DIA. However, 'while the Israelis have agreed to US exploitation of the SA-2 system within their country,' US Chief of Staff General Robert Glass reported to Washington on 15 November that 'there remains little likelihood that they will soon agree to the evacuation of selected system components to the US.'

In Washington, Israeli officials were locked in negotiations with their counterparts from the US State Department to buy F-4 Phantom fighters and other cutting-edge American weaponry that the US had been reluctant to sell to Tel Aviv. In his report, General Carroll suggested that the Israelis were using access to the more high-value hardware they possessed as a bargaining chip in these talks and 'may be purposely engendering these delays in order to await the outcome of the equipment negotiations in this country.'

The political obstacles were eventually overcome, and in January 1968 General Carroll was able to report that 'excellent cooperation has been received from the Israelis.' He went on to state that 'access had been obtained to some 221 line items of Soviet equipment and, with certain exceptions, technical exploitation had been completed to the extent permitted by existing conditions and available facilities.' Amongst the materiel examined, General Carroll revealed, were 'aircraft remnants and various items of aircraft armament,' including the AA-2 Atoll air-to-air missile. The US teams were also permitted to inspect a captured 'Fan Song' radar and other components from an S-75 missile battery.

The importance of Projects MEXPO and Switch Hit was highlighted in his report to Nitze:

This overall exploitation effort is expected to fill many US intelligence and research and development gaps, some of which are directly associated with the Southeast Asia conflict [Vietnam]. To date, much raw data have been gathered which are essential to detailed analysis. Completion of the detailed analysis is expected to provide the US with an assessment of the capability of weapons and other equipment in the current Soviet inventory, and an insight into the Soviet state of the art pertaining to military items exploited to include design criteria, production quality control and research and development philosophy.

General Carroll also assured Nitze that throughout the operation, 'appropriate security precautions have been taken by both countries to ensure that the knowledge of this effort is controlled on a need-to-know basis.'[15]

Project MEXPO and Switch Hit helped cement the 'special relationship' between Israel and the United States. In November 1968, President Lyndon Johnson's administration approved the sale of an initial 50 F-4 Phantoms long desired by the IAF, greatly enhancing its combat capabilities.

The numerous wars that raged throughout the African continent brought further opportunities for US technical personnel to examine Soviet aircraft. One of these was the Ogaden War of 1977-78, a border conflict between Ethiopia and Somalia. In the course of the war several Ethiopian MiGs crashed on Somali territory and in December 1980, following negotiations with the Somali government, permission was granted for a team from the FTD to recover components from the wreckage of an Ethiopian MiG-23BN and MiG-21bis, the most advanced model of the Soviet fighter. 'As the crashes occurred in remote and metal-hungry areas,' reported Lieutenant General Eugene Tighe, Director of the DIA at the time, 'considerable effort on the part of local village chiefs and the Somali authorities was required to retrieve missing pieces but excellent success was achieved.'

Although examination of the MiG-23 'disclosed no surprises', some parts were brought back to the US, including the cockpit section. Of greater intelligence value was the MiG-21bis, 'the first of the type seen' by US analysts, according to Tighe. Both the engine and radar were 'returned to FTD for disassembly and detailed analysis.'

Overall, General Tighe felt that the mission to Somalia was 'an unusually fortuitous and fruitful one.'[16]

Capturing Satan's Chariot

The last major proxy conflict of the Cold War took place in Afghanistan, which pitted the Soviet Army against local guerrilla fighters (known collectively as the *Mujahideen*), who were supported by the US, Pakistan, Britain and several Middle Eastern states.

Concerned about growing political instability and a rural insurgency in the country, fuelled by the autocratic rule of Afghanistan's president, Hafizullah Amin, in December 1979 the Kremlin orchestrated a coup, KGB special forces assassinating Amin and installing a more pliant puppet president in the capital Kabul, Babrak Karmal. But this did nothing to quell the insurgency being waged by the *Mujahideen*, and the Soviets, like the Americans in Vietnam, found themselves sucked into a quagmire. In the decade-long struggle that ensued, over 14,000 Soviets and an estimated one million Afghans lost their lives, which ended in a humiliating Soviet withdrawal.

Also like the Americans in Vietnam, the Soviets came to rely heavily on airpower in Afghanistan, and the war presented Western intelligence agencies with opportunities to acquire the latest Warsaw Pact fighter-bombers and helicopters.

As part of Moscow's policy of handing over responsibility for internal security to locally-recruited government forces, the Soviets helped build up an Afghan Air Force, equipped with Soviet aircraft. But the morale of its crews was poor and their loyalty to the Kabul regime questionable, as a 1985 CIA intelligence report observed: 'The Afghan Air Force remains an unreliable Soviet ally that suffers from both a lack of combat will and internal unrest … Defections also plague the [Afghan] Air Force.'[17] One early such defection occurred in February 1981, when the Afghan pilot of a Mil Mi-8 'Hip' utility helicopter flew to neighbouring Pakistan, a US ally that was helping the CIA coordinate arms supplies to the *Mujahideen*. Two years later, on 20 November 1983, a Sukhoi Su-7BM 'Fitter' fighter-bomber piloted by an Afghan Air Force defector made a crash-landing at the small civil airport in the city of Dalbandin, in Pakistan's Balochistan province.

The most useful aircraft used by Soviet and Afghan government forces in the country was the Mil Mi-24 'Hind' assault gunship. Heavily armed, with a four-barrelled 12.7mm Yak-B machine-gun in a chin turret, and AT-2 anti-tank guided missiles and S-5 55mm rocket pods mounted on its stub wings, it also featured impressive ballistic protection, with titanium armour around the crew compartment and vital components which could deflect rounds up to 12.7mm calibre. Unlike American gunships like the Cobra and Apache, it could also carry up to eight troops. The 'Hind' soon became the most feared aircraft in Afghan skies, the *Mujahideen* calling it *Shaitan-Arba* ('Satan's Chariot').

Although difficult to shoot down, the 'Hind' was not invulnerable, and the *Mujahideen* did occasionally succeed in downing Mi-24s with heavy machine-guns or rocket-propelled grenades. The CIA, constrained from operating inside Afghanistan due to Congressional oversight, turned to SIS for help in a covert operation to recover samples of armour from a crashed 'Hind'. As Margaret Thatcher's government had insisted that British involvement in the Afghan

War remain clandestine and officially deniable, SIS hired a team of mercenaries, which included several former members of the SAS, to undertake the high-risk mission. When exactly the extraction operation took place is uncertain, one account suggesting either 1980 or 1981. The British team was taken to the site of the crashed Mi-24 by their *Mujahideen* guides and used a mechanical saw to remove parts of the helicopter's titanium armour.[18] The samples of armour plating, along with other important parts, were taken back across the border into Pakistan on pack-mules.

Afghan Air Force defectors delivered a pair of Mi-25s (the export variant of the 'Hind', which lacked some of the more sensitive equipment of the Soviet version) to Pakistan in June 1985. CIA officers were soon on the scene, and Pakistan's president, General Zia, loaned one of the Mi-25s to the Americans. Information gained from analysis of the Mi-25 allowed US engineers to modify the FIM-92 Stinger shoulder-launched SAM – supplied in large quantities to the *Mujahideen* by the CIA from 1986 – to overcome the Hind's anti-missile countermeasures.

The American effort to secure an intact 'Hind' of their own did not let up. In 1988, the focus of this effort shifted from Afghanistan to North Africa. After Chad captured the Oaudi Doum air base during its border conflict with Libya in 1987, Chadian troops found a number of abandoned Libyan Air Force aircraft, including three Mi-25Ds. An agreement was reached with the US to exchange one of the Hinds for $2 million and a consignment of Stinger missiles. The recovery operation was named Mount Hope III and launched on the night of 11 June 1988, with two US Chinook helicopters lifting the 'Hind' out of the Oaudi Doum base and taking it back to N'Djamena International airport in Chad, from where it was flown back to the United States on a USAF C-5 Galaxy transporter. After US technicians had examined it in minute detail, the 'Hind' was donated to the Southern Museum of Flight in Alabama for public display.

In the final months of the Soviet war in Afghanistan one more prize fell into American hands. On the night of 4 August 1988, a Sukhoi Su-25 'Frogfoot' ground-attack aircraft on a mission to bomb a *Mujahideen* ammunition dump near the Afghan-Pakistan border was intercepted and shot down by a Pakistani Air Force F-16. The pilot, Colonel Alexander Rutskoi, successfully ejected over Pakistani territory and spent five days on the run before being captured. After high-level negotiations between Moscow and Islamabad, Rutskoi was repatriated two weeks later. Upon being informed by Pakistan's ISI intelligence agency that Rutskoi's Su-25 had been found and many of its components and weapons were largely intact, Milt Bearden, a CIA officer stationed in Islamabad, struck a deal with the ISI for the aircraft wreckage. The price? Eight Toyota Hilux pickup trucks and a few Chinese Type 63 12-barrelled multiple rocket launchers, which would be passed on to the *Mujahideen*. The captured Su-25 was soon on its way back to America aboard a USAF transport aircraft.

Chapter Eleven

Soviet S & T in the Era of Détente

In 1969, the newly elected US president Richard Nixon and his influential National Security Adviser Henry Kissinger formulated the policy of détente, the aim of which was to reduce East-West tensions by improving relations between the United States, China and the Soviet Union after the turbulence of the 1960s. In February 1972, Nixon and Chinese leader Mao Zedong held a historic meeting in Beijing. Three months later, he became the first US president to meet with a Soviet leader in Moscow, when he held talks with Leonid Brezhnev, principally regarding nuclear arms reduction.

But the apparent thaw in East-West relations brought no let up in the Soviets' S & T collection effort in the US. Indeed, as a key element of the policy was the easing of trade restrictions between the two superpowers, détente opened up new opportunities for Directorate T to acquire the latest American technology. Although the US State Department naturally maintained its embargo on the export of goods with military potential to the USSR, there was still much Soviet engineers could learn from America's civil aviation sector, with many of the items purchased by the Soviets proving to be of considerable use to the country's defence industry.

In 1973 a 40-strong Soviet trade delegation toured a Boeing plant in the States to inspect production of their 747 airliner, with a view to buying some of the jumbo jets. But the Soviet visitors had no intention of placing any orders. In a repeat of a ploy successfully used by metallurgist S.T. Kishkin in England in 1946 (see Chapter Four), the CIA later discovered that a member of the Soviet team wore shoes with special absorbent soles designed to pick up metal samples from the factory floor, which were later sent back to Moscow for analysis. The Agency dubbed this tactic 'gumshoe'.[1] These samples reportedly helped in the development of the alloys used in Ilyushin's Il-76 military transport. Moscow even proposed a deal for Lockheed to set up a production facility in the USSR to build civil aircraft, in exchange for the purchase of 50 transport planes. These negotiations came to naught.

One deal that did go through was the sale of microprocessors and semiconductors. This was of special importance to Moscow, as computers and microelectronics were areas where the Americans and West Europeans held a substantial lead over the Soviets. The CIA suspected the technology the Soviets obtained from these items was subsequently diverted to military programmes.

Only years later did the Americans learn the true scale of Soviet exploitation of Western technology obtained through legal means. '[The Soviets] have succeeded in acquiring the most advanced Western technology by using, in part, their scientific and technological agreements with the West to facilitate access to the new technologies that are emerging from the Free World's applied scientific research efforts,' revealed a CIA report from 1982. The report added that Soviet aircraft designers 'were in particular need of data on US technological advances, but more importantly they needed information on aerospace manufacturing techniques.'[2]

Meanwhile, the Soviets pressed on with – and intensified – their campaign of covert aviation technology theft through Directorate T's Line X network in the States, targeting military technology that was subject to strict export controls.

By the 1970s, Directorate T had emerged as one of the most important and successful sections within the KGB's First Chief Directorate (the arm of the security service concerned with foreign espionage), and its prestige within the Soviet Union's gigantic intelligence structure was at its height. Line X officers stationed in the US residencies had succeeded in recruiting sources in some of America's most important defence contractors, including the computer giant IBM, General Electric, McDonnell Douglas, Sperry-Rand and the TRW Corporation (which was involved in the design of US spy satellites), as well as the USAF and DARCOM (the body responsible for R & D of US military equipment). By 1975 it was running 77 agents employed by US companies connected to the defence industry, some of whom were working for subsidiaries outside the United States.

SIGINT also provided the KGB with a rich source of information on American military aviation. In 1966, the KGB began an eavesdropping operation known as *Pochin* (Russian for 'Start'), in which a secret listening post was set up in the Soviet embassy in Washington to intercept confidential communications, including those from the White House and the Pentagon. Over the next few years, the operation expanded with the addition of further listening posts in the Soviet residencies in New York and San Francisco, codenamed, respectively, *Proba* ('Test') and *Vesna* ('Spring'), which intercepted telephone and fax communications from leading US defence contractors, including Grumman, Lockheed, Hughes Aircraft and Fairchild. Throughout the 1970s, these SIGINT stations allowed the Soviets to harvest a wealth of classified data on many military aircraft programmes, such as the F-14, F-15, F-16 and F/A-18 fighters, the A-10 'Thunderbolt II' attack and EF-111 'Raven' electronic countermeasures aircraft.[3] The flow of intel from the *Pochin*, *Proba* and *Vesna* listening posts only began to dry up from 1980, when the US tightened its communications security.

Between 1972 and 1977, Directorate T reportedly provided the VPK and GKNT, the two main customers for its product, with over 140,000 documents relating to foreign research and development, and more than 20,000 samples of material. This was said to have shaved as much as six years off the time spent on R & D, as well as boosting the Soviet economy by over a billion roubles.[4]

The growing power and influence of Directorate T prompted its new chief General Leonid Zaitsev to lobby for the department to be made a separate entity within the KGB. While this proposal was vetoed by Vladimir Kryuchkov, head of the FCD, Directorate T was granted greater autonomy than the other KGB directorates and its officers came to be regarded as an elite within the service.

The FBI strikes back

Directorate T was not, however, immune from failure. In 1970, an engineer working for the Grumman Aerospace Corporation in Long Island fell into conversation at a social function in New York with a Russian, who gave his name as Sergei Petrov and told him he was a trained aeronautical engineer, currently working as a translator at the UN. Petrov commiserated with the American as he confided his fears that, with Grumman in the process of downsizing, he might fall victim to job cuts at the firm. They arranged to meet again a week later at a restaurant in Long Island, where over a steak dinner Petrov told his companion he was working on a doctoral thesis on aerospace engineering and would appreciate any data he could provide regarding Grumman's new F-14 carrier fighter.

Developed as the next generation of air defence interceptor for the US Navy's carrier fleet, the F-14 Tomcat was powered by two Pratt & Whitney TF30 engines, had a range of 3,000km, and was armed with the extremely long-range AIM-54 Phoenix air-to-air missile. Making its maiden flight on 21 December 1970, the F-14 represented a major advance in capabilities over its predecessors, the F-8 Crusader and F-4 Phantom.

Petrov assured the American he would be willing to pay him generously for any information on the F-14, and particularly its swing-wing mechanism. The engineer promised Petrov he would consider the offer and they arranged to meet again in three weeks' time. Strongly suspecting that his new Russian friend was a spy trying to recruit him as a source, the engineer reported the approach to Grumman's security department, which in turn informed the FBI. Agents from the Bureau's New York Field Office persuaded the engineer to continue meeting Petrov, whom they placed under surveillance. At these meetings, held at monthly intervals at different restaurants in Long Island, Petrov pressed him for more information on the F-14, offering sums of between $100 and $300, depending on how valuable he judged the material.

The engineer supplied Petrov with routine reports on the fighter which had first been cleared by his FBI handlers. But the Russian became impatient for more confidential material. When the engineer protested that this would be difficult to obtain in his current position, Petrov encouraged him to seek a transfer to another role within the company where he would have access to such documents. To placate Petrov, the FBI allowed the engineer to hand over copies of technical drawings of the F-14's swing-wing design in March 1971, made using a portable copying machine provided by Petrov.

In May 1971 Petrov told the engineer he had to return to the Soviet Union but that their meetings would resume upon his return to the States in August, and that he should continue copying documents in his absence. At another meeting in November that year, Petrov supplied the engineer with a 35mm camera and high-strength lamp for photographing documents, advising him to smuggle the rolls of film out of the Grumman premises in a cigarette packet.

By now, the FBI had all the evidence they needed against Petrov and decided the time had come to make their move. After meeting in a restaurant in the town of Brookhaven in Long Island on 14 February 1972, Petrov and the engineer left together and got into the latter's car, where Petrov handed over an envelope of cash in exchange for a roll of film of photographed pages from an F-14 engineering report. As Petrov was walking away towards the car park's exit, waiting FBI agents arrested him with the incriminating material handed over by the engineer.

Charged with espionage and violation of the Foreign Agents Registration Act, Petrov was arraigned and released on $100,000 bail. As a high-profile Soviet spy trial risked jeopardizing the closer relations with the USSR the Nixon administration was determined to foster, six months later the charges against Petrov were dropped, following White House intervention, and he was quietly deported to the Soviet Union.

Despite this setback, the Soviets remained keen to acquire intelligence on the F-14, which was correctly judged as a serious threat to Soviet naval aviation's long-range maritime patrol aircraft and cruise missile-armed anti-ship bombers, like the Tu-142 'Bear' and Tu-22 'Backfire'.

Of special interest to the Soviets were the F-14's AN/AWG-9 radar, which could track up to 24 targets simultaneously, and the AIM-54 Phoenix it carried. Developed by Hughes Aircraft, the Phoenix was a radar-guided BVR (Beyond Visual Range) air-to-air missile, four metres long, with a range of around 135km. The F-14 could carry up to six of these missiles.

Another opportunity to gain access to the secrets of the F-14 and the Phoenix presented itself in 1976. On the afternoon of 14 September, the carrier USS *John F. Kennedy* was taking part in the major naval exercise Teamwork 1976 in the north Atlantic, when one of her F-14s hurtled off the flight deck due to a malfunctioning engine control system. The crew, pilot Lieutenant John Kosich and Radar Intercept Officer Lieutenant Louis Seymour, successfully ejected and landed on the flight deck, Seymour suffering minor injuries. But the F-14 crashed into the sea and disappeared beneath the water.

The naval exercises were being closely monitored by a Soviet Tu-142, a *Kresta II*-class missile cruiser and a spy trawler, and the Americans feared the Soviets would try to salvage the F-14. And so a race began to recover the wreckage. A US Navy official told the press soon after the crash that 'we would just as soon not leave the plane lying around on the bottom for [the Soviets] to pick up'.[5] First, though, the Americans had to locate the Tomcat. The salvage tug *Shakori* and the nuclear-powered research and recovery mini submarine *NR-1* led the operation.

It was no easy task. As the captain of the *NR-1*, Commander Holifield, put it, finding the F-14 was like 'looking for a needle in a grassy front yard with only the aid of a penlight.'[6] Gale force winds and 20-foot waves added to the search team's difficulties.

Finally, on 15 October the *Shakori* located the F-14, around 60 miles northwest of the Orkneys, at a depth of almost 2,000ft. The *NR-1* inspected the wreck and found evidence to suggest that the Soviets had discovered it before them and had made an unsuccessful attempt to recover parts of the aircraft. More worrying still, the Phoenix missile was missing from its weapons station. Had the Soviets succeeded in retrieving it from the seabed without the Americans realizing? With no way of knowing, the hunt for the missile continued. Two weeks later, the US Navy's fears were dispelled when the *NR-1* discovered the missile lying near the wreck site, largely intact, and brought it to the surface with the assistance of the submarine rescue ship USS *Sunbird*. The F-14, meanwhile, was towed by salvage tug to a sheltered anchorage in the Orkneys and also brought to the surface, on 11 November. The salvage operation cost $1.5 million.

The secrets of the AN/AWG-9 radar and Phoenix missile would be uncovered by the communists soon after, however. Not by the Soviets, but by their Polish allies, through a source recruited in the defence contractor Hughes Aircraft, manufacturer of the Phoenix.

William Holden Bell was a 58-year-old project manager in Hughes' Radar Systems Group at its El Segundo facility, who had worked for the company since 1952. A divorcee who had recently remarried to a much younger woman, he was suffering financial difficulties and so was vulnerable to recruitment by a foreign intelligence agency. In 1977 he was befriended by 26-year-old Marian Zacharski, a Polish neighbour in the apartment complex in Playa del Rey, California, where he lived. Zacharski's official cover in the US was that he was a senior representative of POLAMCO, a tool machine company. Unofficially, he was an operative for the SB, Poland's state intelligence service.

Learning of his neighbour's money woes, Zacharski offered to pay him for classified information on Hughes' products. Bell agreed and, using a camera provided by the Pole, photographed documents on the Phoenix, Hawk and Patriot missiles, as well as the radar systems of the F-14, F-15, the Rockwell B1-B strategic bomber and the top-secret Lockheed F-117 Nighthawk stealth bomber. Most of this material, for which he was paid at least $110,000, in cash and gold coins, was handed over by Bell during trips to Europe.

The leak of classified documents to the SB only ended in 1981, after the FBI was tipped off about Zacharski's status as a spy by a double agent inside the SB, Jerzy Korycinski (codename 'Caribou'), who had been recruited by the CIA. On 24 June, Bell was visited by the FBI, confessed his treason, and agreed to cooperate in their investigation into Zacharski in return for a reduced sentence. When he next met with his handler four days later, Bell was wearing a wire and recorded Zacharski requesting more information on the F-15's radar. The FBI then swooped, arresting both men. After a two-day trial in December 1981, Zacharski was found guilty of espionage and given a life term, while Bell, who had pleaded guilty, received eight years. Four years later, Zacharski and three other Soviet bloc spies captured in America were exchanged in a prisoner swap at the Glienicke Bridge.

After the fall of communism in Poland, Zacharski became a successful businessman.

Cases such as William Bell's prompted Edward O'Malley, assistant director for intelligence at the FBI, to warn in 1981 that 'the threat [from Soviet bloc espionage] has increased substantially.'[7]

Directorate T operations in Europe

Although the United States was by far the most important target for Soviet S & T operations, accounting for over sixty per cent of all intelligence gathered during this period, Directorate T did not neglect Western Europe. The CIA regarded Western Europe as a generally more benign environment for Soviet technological espionage than the US. 'The acquisition of Western technology operationally was assigned the highest priority for collection by local residencies in key West European countries because of the relatively easy access to much US and Western technology in Europe,' an Agency report revealed.[8] Nevertheless, Directorate T's performance in the major European aircraft-producing nations – the UK, France, West Germany and Italy – in the 1970s was decidedly mixed.

Britain had long been an important source for S & T. But Directorate T's British operations suffered a serious setback in 1971. Warned by MI5 that the sheer number of KGB officers operating in the country under diplomatic cover was stretching its limited resources to breaking point, in the summer of 1971 the British Government drew up plans for a mass expulsion of Soviet 'diplomats'. The plan, codenamed Operation Foot, was implemented on 24 September when the Soviet *chargé d'affaires* Ivan Ipolitov was told by the Permanent Under-Secretary at the Foreign Office, Sir Denis Greenhill, that 90 Soviet diplomats were to be declared *personae non gratae* and expelled, while a further 15 normally based in the UK but currently out of the country would be barred from re-entry. All of these diplomats were KGB and GRU officers, identified as such by a recent defector from the Soviet embassy.

Oleg Gordievsky, a KGB officer who worked as a double agent for SIS, described Operation Foot as a 'bombshell', which 'shocked [Moscow] Centre profoundly.' Oleg Kalugin, a senior officer in the KGB at the time, considered it a setback from which the KGB never fully recovered.[9] The effects of Operation Foot were felt especially keenly by Directorate T. Several of those expelled were Line X officers, among them Lev Shertsnev, the Soviet embassy's first secretary, who had been head of Line X operations in the UK since 1968.

It took several years before Line X was able to reactivate some of their sources in Britain, including 'Ace', their long-serving spy in British European Airways. But S & T operations in Britain would never reach pre-Foot levels. From the top European intelligence target, by the end of the 1970s the UK had slipped behind both France and West Germany, accounting for only 7.5 per cent of the S & T material accumulated by Directorate T.

By then, the main source of S & T in Europe was West Germany, accounting for over 10.5 per cent of intelligence gathered. In July 1977, KGB chairman Vladimir Kryuchkov identified West Germany as 'the most important component of the [NATO] military bloc' in Europe, and stated that as 'military scientific research studies in the fields of atomic energy, aviation, rocket construction, electronics, chemistry and biology are being intensively pursued' there, intelligence-gathering in West Germany should be accorded a high priority.[10]

The most important German aviation spy handled by Directorate T during the latter part of the Cold War, and certainly the longest-serving, was Manfred Rotsch, codename 'Emil'. Born in 1924, Rotsch studied engineering at the Technical University of Dresden from 1948–52. Believed to have been recruited by the KGB in 1954, in May 1955 he was sent into West Germany posing as a political refugee.

Soon after, he gained employment with the Junkers aircraft company (which was later absorbed by Messerschmitt), and passed on details to the KGB of the French-designed Fouga Magister trainer and light attack jet, manufactured under licence by Messerschmitt. In 1969, an amalgamation of Messerschmitt and the engineering companies Bölkow and Blohm created MBB, West Germany's largest aerospace company, and Rotsch became head of its Planning Department.

One of the projects he was involved in was the Panavia Tornado MRCA (Multi-Role Combat Aircraft), the most important European military aircraft programme of the late Cold War period. An Anglo–German–Italian strike aircraft developed by a multinational consortium comprising the British Aircraft Corporation, MBB and Fiat, the Tornado project began life in 1968 as a replacement for the ageing Lockheed F-104 Starfighter in Luftwaffe, *Marineflieger* (West German naval aviation) and *Aeronautica Militare Italiana* (Italian Air Force) service, and to fill an RAF requirement for a low-level precision strike and interdiction bomber. The prototype first flew in August 1974 and began entering squadron service with the air arms of the three partner nations from 1982.

Truly earning its 'multi-role' designation, the Tornado would eventually serve in the strike – including, in the RAF's case, tactical nuclear strike – battlefield interdiction, reconnaissance, anti-ship attack and ECM roles, while the British also developed a dedicated interceptor variant.

Throughout the development of the Tornado, Rotsch kept the KGB fully informed of the project's progress, meeting with his handler at least five times a year and also sending information in microdot messages. In 1984, after Rotsch's unmasking as a spy, an unnamed intelligence expert told the West German newspaper *Bild* that thanks to his treachery the Soviets 'know everything about the aircraft.'[11] Supplementing the information on the Tornado provided by agent 'Emil' was intel from an Italian source. Though trailing in last among the 'big four' of West European states as a source of S & T intelligence, some very useful Italian sources were recruited by Directorate T. One of these was a university lecturer given the apt codename '*Uchitel*' ('Teacher'), who had numerous contacts in Italian industry and research facilities, including Fiat.

Second only to West Germany for S & T in Europe was France. By 1975, the KGB was running 55 agents in the country and their Paris residency was staffed by 22 Line X officers, more than double the number stationed in London. One of Directorate T's prize assets in the 1970s was 'Alan', who worked for a defence contractor and from 1972 supplied documents on advanced military technology, including missile systems and night-vision gear for helicopter crews. His spying career came to an end in 1978 after he was fired by his employer on suspicion of carrying out industrial espionage on behalf of a Western intelligence agency. Like most informants recruited during this time, 'Alan' spied for financial gain – he was reportedly paid over 200,000 francs a year by the KGB.

Another valuable French Line X source of the 1970s was 'Laurent', of whom little is known, other than he was a scientist working in the aerospace industry and was highly regarded by the KGB, being included in a list of 13 agents deemed to be of particular value.

The decade of détente, which came to an end with the Soviet invasion of Afghanistan in December 1979, sending relations between the West and the USSR back into the deep freeze, had been a productive time for Soviet S & T gathering. 'Since 1970, Line X had obtained thousands of documents and sample products, in such quantity that it appeared that the Soviet military and civil sectors were in large measure running their research on that of the West, particularly the United States. Our science was supporting their national defense,' insisted the US security adviser Gus Weiss.[12]

But as the Cold War entered its final decade, the West hit back with a major intelligence coup that was to have far-reaching implications for Directorate T.

Chapter Twelve

The Final Years

By the late 1970s, the complacency within US government circles regarding the threat posed by Soviet technology theft was beginning to give way to a more realistic and hard-headed assessment of the problem. Surprisingly, this occurred during the presidency of Jimmy Carter, generally regarded as a Cold War 'dove' anxious to improve East-West relations. In August 1977 he issued Presidential Review Memorandum 31 instructing the Security Council to provide an assessment of the 'military, political and economic implications for the US and its allies of technology transfer to the various Communist states.'[1] Washington's attitude towards the Soviet Union hardened further under Ronald Reagan, elected president in 1980 on a pledge to confront the 'evil empire', as he famously called the USSR.

In April 1982, the CIA compiled the first detailed intelligence report covering the whole area of Soviet technology theft. Entitled *Soviet Acquisition of Western Technology*, the report revealed the frightening scale of Soviet espionage activity targeting the West's defence industries. 'Today this Soviet effort is massive, well planned, and well managed – a national-level program approved at the highest [Communist] Party and governmental levels.' It continued:

The Soviet intelligence services have the primary responsibility for collecting Western classified, export controlled, and proprietary technology, using both clandestine and overt collection methods. They in turn make extensive use of many of the East European Intelligence services for their efforts in acquiring Western technology. These countries are paid in part with Soviet military equipment and weapons . . . These intelligence organizations have been so successful at acquiring Western technology that the manpower levels they allocate to this effort have increased significantly since the 1970s to the point where there are now several thousand technology collection officers at work.[2]

Military and civil aviation was highlighted in the report as a top target for these intelligence collectors. However, 'while some of the Soviet acquisition in the aircraft area appears directed toward the development of countermeasures against Western systems, the Soviets appear to target data on Western aircraft primarily to acquire the technology' in order to incorporate this technology into their own projects.

Assimilation of Western technology has been of great benefit to both their military and commercial aircraft development programs – to the extent that aircraft from certain Soviet military design bureaus are to a significant degree copies of aircraft of Western design. Soviet military aircraft designers have 'ordered' documents on Western aircraft and received them within a few months.[3]

One example cited in the report was the Lockheed C-5A Galaxy, a heavy-lift strategic military transport aircraft, plans and drawings of which were obtained 'early in its development cycle.' Catapult technology for the Soviet Navy's aircraft carriers then under construction was also acquired by their intelligence services, the report disclosed.

S & T intelligence collection had allowed the Soviets to 'modernize critical sectors of their military industry and reduce engineering risks by following or copying proven Western designs, thereby limiting the rise in their military production costs.' It had also helped them 'achieve greater weapons performance than if they had to rely solely on their own technology' and 'incorporate countermeasures to Western weapons early in the development of their own weapons programs.' This intelligence effort had saved the Soviets 'millions of dollars in R & D costs, and years in R & D development lead time.'[4]

In a later update to the report, the CIA correctly identified the VPK as the agency that coordinates the Soviet campaign of technology theft, and included a detailed breakdown of the Soviet S & T structure:

> The VPK is the most powerful organization in the defense-research establishment, comprising the top executives of the key defense manufacturing ministries.
>
> To satisfy these requirements, the VPK controls a national fund, amounting to some half a billion rubles each year. Once approved by the VPK, requirements are selectively levied among the KGB, the GRU, and at least four other national-level collection agencies, as well as surrogates among the East European intelligence services.
>
> The State Committee for Science and Technology (GKNT) acts as a collector and as the central processor for the national-level program. It also monitors the absorption and assimilation of Western technology by the defense ministries.[5]

The Soviet campaign of technology theft was likely to expand during the 1980s, the CIA warned, and predicted that in the aviation field they would prioritize intelligence on composite materials to reduce aircraft weight, turbofan engines for strategic transports, inertial navigation systems, and CAD (Computer-Aided Design) technology ('an area in which the Soviets are clearly impressed by US progress,' the CIA stated).

The challenge of stemming this flow of Western technological secrets to the Soviet bloc was 'formidable', the report admitted, and 'likely to become even more difficult in the future' as the Soviet appetite for such technology would remain 'voracious'. The report highlighted the scale of the problem by pointing out that in the United States alone there were over 11,000 defence contractors, as well as hundreds more with overseas subsidiaries. This meant that the resources of US counterintelligence were stretched thin. Their counterparts in the Western European intelligence agencies were also 'feeling the burden of well-orchestrated Soviet Bloc [S & T espionage] efforts.'

In response to this threat, 'the West,' argued the report's authors, 'will need to organize more effectively than it has in the past to protect its military, industrial, commercial, and scientific communities.'[6]

The Farewell Dossier

The information contained in this influential report was remarkably accurate and detailed. Which is hardly surprising, given that it was based largely on intel provided by a mole at the very heart of Directorate T.

Vladimir Vetrov was born in Moscow in 1932 and studied mechanical engineering at the Bauman Higher Technical School. In 1959 he joined the KGB and was assigned to the foreign relations department of GKET (the State Committee of Electronic Technology), where his job was to develop contacts among the foreign engineers who visited the USSR. In 1965 he was posted to the Soviet residency in Paris, attached to the Ministry of Foreign Trade under diplomatic cover. Officially, his job was to negotiate trade deals with French engineering firms, but his real role was to recruit sources in French industry. One of the most important sources he recruited during his time in France was Pierre Bourdiol (codename 'Karl'), who worked on aerospace projects for the electronics company Thomson-CSF.

After his return to Moscow in 1970 he joined Directorate T, and in 1974 he was sent to Montreal as chief engineer of the Soviet trade mission. But less than a year into his Canadian posting he was suddenly recalled to Moscow, apparently after having been identified as an intelligence officer by the RCMP. With his cover blown, he would never again be given a coveted overseas posting and instead was moved to a bureaucratic position in Moscow as an assistant to the head of Directorate T's Fourth Department (Information and Analysis).

By the beginning of the 1980s, Vetrov was in a state of personal and professional crisis. His marriage was on the rocks, he was becoming increasingly dependent on alcohol, and frustration over his lack of career advancement had developed into an intense resentment towards the KGB, on which he now sought to inflict maximum damage by offering to spy for the French domestic intelligence service, the DST.

In early 1981 he made contact with the agency via an acquaintance from his days in Paris, Jacques Prevost, a Thomson-CSF executive with ties to the DST. The recruitment of Vetrov was a major coup for the French, who gave him the codename 'Farewell', and he began passing on masses of intelligence on Directorate T and its operations to his French handlers in Moscow, Xavier Amell and Patrick Ferrant, which became known as 'the Farewell Dossier'.

At a G7 summit held in Ottawa in July 1981, French President Francois Mitterand brought Ronald Reagan into the 'Farewell' secret, as so much of the intel Vetrov was providing – around 70 per cent – related to Directorate T activities in the US (and also to impress the Americans with what was a stunning coup for French intelligence in recruiting Vetrov in the first place). Reagan was indeed suitably impressed, commenting that 'Farewell' was the 'biggest fish' recruited by the allies since the Second World War.[7]

The scale of Soviet S & T espionage revealed by Agent 'Farewell's' disclosures came as a shock to the US intelligence community. The CIA reported that 'the magnitude of the Soviets' collection effort and their ability to assimilate collected equipment and technology are far greater than was previously believed.' Another US intelligence report revealed that the secrets of the Rockwell B-1B bomber, the McDonnell Douglas F-15 Strike Eagle and Boeing's new F/A-18 Hornet carrier fighter had all been compromised. Classified material gathered by Directorate T on the F/A-18's AN/APG-65 radar had proved especially useful to the Soviets, the report explained. 'The documentation of the F-18 fire-control radar served as the technical basis for new lookdown/shootdown engagement radars for the latest generation of Soviet fighters.' This material, the report added, had saved the Soviets five years in development time and 35 million roubles.[8] It was also reported that S & T intelligence gained in the West had contributed to the development of more than 50 per cent of all Soviet military projects by the end of the 1970s.

In August 1981, the chief of the DST, Marcel Chalet, travelled to Washington to give a full briefing on the 'Farewell' case to the US Vice-President George Bush, who was left 'clearly shaken' by the extent of Soviet knowledge of the US defence network. A few months later, the DST shared the secrets of the Farewell Dossier with the heads of the intelligence services of France's main European allies, the UK, West Germany and Italy.

Meanwhile, Vetrov's life was rapidly spinning out of control. His alcoholism was growing worse and he was becoming increasingly paranoid. Vetrov's career as a double agent came to a sudden and violent end on the evening of 22 February 1982. Meeting with his mistress Ludmila Ochikina, a translator at Directorate T, in his car parked near the KGB's headquarters, an argument broke out and Vetrov stabbed her several times. A retired policeman who intervened was also stabbed by Vetrov, fatally. Ludmila, though badly injured, managed to escape from Vetrov and raised the alarm.

Vetrov was charged with murder and attempted murder and held in Lefortovo prison. After an eight-day trial at the Moscow military regional court, on 3 November 1982 he was found guilty and sentenced to 15 years. Only months later did the KGB learn that he had committed treason as well as murder, though it remains unclear how they came to suspect him.[9] In September 1983 he confessed to spying for the French to his KGB interrogators. Convicted of high treason at the USSR's Supreme Court in December 1984, he was sentenced to death and executed the following month.

But the damage to the KGB in general and Directorate T in particular had already been done, and it was extensive; Vetrov had provided the DST with almost 3,000 pages of classified documents and the names of 422 KGB officers, around 250 of whom were Line X officers stationed abroad, along with at least 57 of their assets.[10]

In the US, the CIA formed a dedicated unit, the Technology Transfer Intelligence Center, to assess the information in the Farewell Dossier and overhaul security to plug the leak of America's technological secrets. A similar task group was also set up by the Pentagon.

In April 1983 Mitterand ordered the mass expulsion of 47 Soviet intelligence officers working under diplomatic cover in France (many of them attached to Directorate T) who had been identified in the Farewell Dossier, ostensibly in response to the discovery that the KGB had bugged the teleprinters in the French embassy in Moscow. Seven months later, Pierre Bourdiol, the aerospace engineer recruited by Vetrov himself, was arrested and later received a five-year prison sentence.

One of Directorate T's sources betrayed by Vetrov was the German engineer Manfred Rotsch. The BfV launched an investigation into Rotsch, who had given the KGB full details of the Tornado strike aircraft, as well as information on the Milan anti-tank and Komoran anti-ship missiles. He was eventually arrested in September 1984, found guilty of espionage at his trial two years later in Munich and sentenced to eight and a half years. After serving just a few months he was exchanged in a prisoner swap for Christa-Klein Schumann, who had been jailed in the GDR in 1980 for alleged spying activities. Rotsch, who had grown attached to his life in the West, returned to live in Munich after the fall of the Berlin Wall.

As a direct result of information in the Farewell Dossier, a total of 148 Soviet bloc intelligence officers were expelled from various residencies around the world.[11]

Thomas Reed, a Reagan policy adviser, claimed that, thanks to the Farewell Dossier, the US and its allies were able to roll up the entire Line X network.[12] The US intelligence adviser Gus Weiss, meanwhile, stated that 'the heart of Soviet technology collection crumbled and would not recover'.[13]

The technology gap

Vetrov's treachery was unquestionably a hammer-blow to Soviet S & T collection, at least in the short term. But Directorate T did recover, to some extent, and resumed its campaign of industrial espionage. By the mid-1980s, the CIA were reporting that the Soviets 'continue to rely heavily on acquisition of Western technology', while the Agency's director William Casey warned that 'gaining access to our advanced technology continues to be their top priority.'[14] Shortly before his appointment as General Secretary of the USSR, Mikhail Gorbachev visited the Soviet embassy in London and praised the work of Directorate T's Line X officers.

The greatest challenge facing the Soviets during the last years of the Cold War was not so much acquiring S & T intelligence, but in effectively exploiting it. This was not a new problem. As far back as the 1960s the Politburo was voicing concern that the country's industries were failing to make expeditious and efficient use of the S & T being collected in the West. As aviation technology became more sophisticated, the Soviet aviation industry struggled to keep up.

This was largely due to the bureaucratic nature of the communist system, which stifled innovation and discouraged personal initiative.

A prime example of this was the Tupolev Tu-144. A supersonic airliner conceived as the Soviet answer to the Anglo-French Concorde (designed and built by the British Aircraft Corporation and France's Süd Aviation), the Tu-144 project relied heavily on technology stolen from its European rival. Information on Concorde's Rolls-Royce Olympus 593 engines was supplied by 'Ace', Directorate T's spy in British European Airways, while further technical details and plans were obtained by agents infiltrated into Süd Aviation. The Tu-144 made its maiden flight on New Year's Eve 1968, two months before Concorde. When it was first seen in the West at the Paris Air Show in June 1971, the Tu-144's remarkable resemblance to Concorde prompted the Western press to dub the aircraft 'Concordski'.

But the similarities were largely skin deep. The Tu-144 was less technically advanced than Concorde; its Kuznetsov NK-44 engines were much less fuel efficient than the Olympus units, and its electronics were quite primitive by comparison with those of its European cousin. The Tu-144 was also plagued with technical problems, leading to several fatal crashes. On 3 June 1973 an early production aircraft broke up during a demonstration flight from Le Bourget airport, with the loss of all six of its crew. Limited commercial service commenced with passenger flights from 1977, but these were suspended a few months later following a second fatal crash of a development aircraft on a test flight near Yegoryevsk, 70 miles southeast of Moscow, on 23 May 1978, killing two crew members. This effectively signalled the end of the short and rather inglorious career of the Tu-144. In an effort to salvage the programme, the Tupolev OKB proposed converting some of the airframes for military roles, including missile carrier and long-range maritime reconnaissance. But this proposal was rejected by the commander of the DA, Vasily Reshetnikov, who considered the

Tu-144 entirely unsuitable for military use. Of the 20 built, most were scrapped, although a few airframes were retained for research flights at the LII and for display at aviation museums.

The failure of the Tu-144 was highlighted as an example of the technical deficiencies of the Soviet aviation industry and its inability to effectively incorporate stolen S & T into its own projects in a CIA report from December 1984. 'Western technical secrets frequently could not be exploited because of inadequate Soviet material and technique as well as inability of Soviet engineers to decipher Western technical data. This,' the report noted, 'undercut extensive Soviet technical espionage.'[15]

By the 1980s, the problem had become even more acute, as the technology gap between the West and the USSR threatened to widen into a chasm. Nowhere was the technology gap more evident than in the emerging area of stealth technology, which promised to be the most significant breakthrough in military aviation since the introduction of the jet engine.

Stealth – or Low Observable (LO) – technology is the means of reducing an aircraft's RCS (Radar Cross-Section) to the point of making it effectively invisible to radar. This is done through complex design features intended to deflect or minimize radar returns and the incorporation of Radar Absorbent Materials, or RAM.

The origins of stealth go back to the Second World War. In 1943 Luftwaffe pilots Reimar and Walter Horton designed the revolutionary Ho 229 fighter/bomber. A 'flying wing' design, the Ho 229 was found to have a very small RCS during flight tests. This was almost certainly a happy accident, rather than intentional.

Although the SR-71 Blackbird spy plane, which entered service in 1966, incorporated some 'stealthy' features, the first true stealth aircraft designed from the outset to be all but invisible to radar was the Lockheed F-117 Nighthawk. Ironically, one of the key figures involved in the early development of the F-117 project, radar specialist

Denys Overholser, was influenced by the groundbreaking theoretical work on stealth carried out by the Soviet engineer and physicist Pytor Ufimtsev, whose research into the reflection of electromagnetic waves was published in a 1962 scientific paper.

The F-117 project began in 1975, when DARPA set Lockheed and Northrop Aviation the challenge of designing a stealth aircraft, known as the XST (Experimental Survivable Testbed). Lockheed's submission won out and two proof-of-concept prototypes were ordered, which went by the codename Have Blue. Shaped like an arrowhead, and featuring highly angled surfaces, the first Have Blue demonstrator flew on 1 December 1977 and showed sufficient promise for the USAF to commission a fully-fledged combat aircraft, developed under conditions of intense secrecy under the codename Project Senior Trend. In 1978 production was approved of what was now called the F-117 Nighthawk (and often referred to by crews as simply 'the Black Jet', due both to its colour and its origins as a highly secret 'black' project).

The F-117 made its maiden flight on 18 June 1981 from Groom Lake and the first of 64 built entered USAF service with the 4450th Tactical Group in late 1983, based at the Tonopah test range in Nevada. But the existence of this radical new warplane was not revealed to the public until 1988, a year before the Nighthawk carried out its first operational mission (bombing an army barracks in Panama during Operation Just Cause).

The veil of secrecy cloaking the Have Blue and F-117 programmes was lifted a little at a press conference on 22 August 1980, when the US Defense Secretary Harold Brown confirmed that the US was developing stealth technology and that 'we have demonstrated to our satisfaction that the technology works.'[16] This public revelation dismayed many US defence officials. The secrets of stealth then inevitably became a prime S & T collection target for Directorate T. Initially, however, the CIA's knowledge of the state of Soviet development of stealth technology was virtually non-existent, forcing the Agency to rely on guesswork, as one report acknowledged: 'Because of the obvious high US interest in this area, the Soviets probably began an intensified research effort in the early 1980s which may have led to a developmental program now under way. Such a program could be well along before we became aware of it.'[17]

In 1984, Thomas Cavanagh, a 40-year-old engineer working at Northrop Grumman's Pico Rivera facility in Los Angeles, contacted the Soviet embassy in Washington and their consulate in San Francisco by payphone, offering to sell classified material on the stealth technology being developed by Northrop for its B-2 Spirit bomber programme. A meeting was arranged at an LA bar on 10 December with two Soviet officials, at which Cavanagh asked for $25,000 for the documents on the B-2. He assured the Soviets this material would be worth 'billions of dollars' to them, and that he could provide more classified information on the B-2 once he had paid off his debts, which were holding up his promotion within the company. Two days later he handed over blueprints and other documents on the B-2. At the third meeting on 18 December, his Soviet contacts paid him the $25,000. The 'Soviets' then revealed themselves to be undercover FBI agents, David Silva and Daniel McLaughlin (both of whom were fluent Russian speakers), who arrested him for espionage. The FBI, it later transpired, had intercepted Cavanagh's phone calls to the Soviet embassy and consulate, allowing the Bureau to set up a sting operation. At his trial in May 1985, Cavanagh pleaded guilty and was sentenced to life imprisonment. FBI director William Webster said that, had Cavanagh succeeded in passing on the documents on the B-2 to the Soviets, it would have caused 'irreparable damage to national security.'[18]

Another case of Soviet stealth espionage involved Subrahmanyam Kota, who ran a computing consultancy that worked on classified US defence projects, and Aluru Prasad, a successful Indian businessman.

From 1985, Kota and Prasad passed on top secret documents relating to military projects to Soviet agents at meetings held in various overseas locations, including Bermuda, Portugal, Switzerland and Cyprus, over a period of several years. In 1989, they obtained information on RAM from an employee of a US defence contractor working on the B-2 project, for which they were paid $100,000 by the KGB. It wasn't until October 1995 that Kota and Prasad were finally arrested by the FBI. Put on trial in June 1996, Kota received a reduced sentence of one year under house arrest and a $50,000 fine after agreeing to testify against his co-accused. Following a mistrial, Prasad was put on trial again in December 1996, found guilty of espionage, and sentenced to 15 months. In October 1996 the men's alleged handler Vladimir Galkin was arrested at New York's JFK airport but released without charge 17 days later following diplomatic intervention.

By 1985, the CIA had become convinced the Soviets were pursuing their own stealth programme, largely based on stolen US research, although the status and details of this programme were, they admitted, unknown. 'Soviet scientists,' the CIA reported, 'have shown an interest in signature-reduction technologies applicable to a broad cross section of aerodynamic vehicles. They have investigated radar-absorbing paints and materials for several years and have acquired technical information, manufacturing equipment, and materials from several foreign sources.'

Soviet industry, however, was at 'a level of technology several years behind that of the United States', and the analysts assessed the prospects of the Soviets fielding an operational manned stealth warplane before the end of the century as slight. 'Developing the technologies required by stealth vehicles will tax the Soviets, even with foreign assistance, but production of such vehicles may be an even more formidable task,' they concluded.[19]

Besides the technical limitations of their aviation industry, efforts to produce a Soviet equivalent of the F-117 or B-2 were further hampered by a 'disinformation' operation put into effect by the CIA. In 1982, the intelligence adviser Gus Weiss proposed to CIA director William Casey that they leak faulty intel on stealth and other defence projects, including the much-vaunted SDI (Strategic Defense Initiative) – referred to at the time by the media as the 'Star Wars' programme – to KGB Line X officers in the US. 'The Pentagon introduced misleading information pertinent to stealth aircraft, space defense, and tactical aircraft,' Weiss confirmed. 'The program had great success, and it was never detected.'

Ultimately, the Soviet Union failed to produce their own stealth aircraft before the end of the Cold War.[20]

Agent CKSphere

During the latter part of the Cold War, the CIA ran a source in the Soviet defence industry whose importance was arguably just as great as that of the DST spy Vladimir Vetrov.

In January 1977, while filling his diplomatic car at a Moscow petrol station, the CIA's station chief Robert Fulton was approached by a stranger who handed him a note before walking off again. The note requested a meeting with an American official to discuss matters of a 'strictly confidential' nature. Suspecting that it might be a KGB trap, the Americans declined to respond to this and several more approaches to US embassy workers made by the same man over the following months.

The CIA only finally took notice when, in December 1977, a US embassy employee passed on an envelope he had received from the man at a Moscow market to the new station chief Gardner 'Gus' Hathaway. Inside were two typewritten pages offering the CIA information on Soviet airborne radars, in particular in the field of 'look-down/shoot-down' radar.

Look-down/shoot-down (or LD/SD) radar enabled a fighter to track and engage low-flying targets which could not be picked out amongst ground clutter on the older generation of A.I. radar. The US began developing LD/SD technology in the 1950s and produced the first set, the AN/ASG-18, in 1960. The first fighter to enter operational service incorporating a look-down/shoot-down capability was the Phantom F-4J model, which was equipped with the Westinghouse AN/AWG-10 fire control system and APG-59 radar.

The Soviets, however, trailed well behind the US in this technology. Their first A.I. radar with a limited LD/SD capability was the Sapfir RP-23 (NATO reporting name 'High Lark II'), fitted to the MiG-23 from 1975 and reportedly based on an AWG-10 set salvaged from an F-4J shot down over North Vietnam in 1967, but already obsolescent when it entered service.[21] With NATO's bomber forces having adopted low-level penetration tactics, and the US introducing a new generation of terrain-hugging cruise missiles like the AGM-86, the lack of an effective LD/SD capability for its interceptors left a major gap in Soviet air defences. So the prospect of obtaining up-to-date intel on the state of Soviet development in this vital field was of the greatest importance, as a CIA memo made clear:

We know that the Soviets do not have a particularly effective look-down, shoot-down radar and that they are working very hard to solve this problem. An effective look-down shoot-down [capability]would pose a serious threat to both the B-52 bomber and the cruise missile, and information on Soviet state-of-the-art in this field is responsive to very high priority intelligence requirements. His offer to provide schematics and sketches of current systems would be of considerable assistance to the analysts.[22]

The CIA gave the would-be informant the codename 'CKSphere'. However, the Agency was still concerned that he could be a KGB 'dangle' – a phoney informant trying to catch Western intelligence officers in the act of carrying out espionage or feed them false intel – and debate raged within the CIA as to whether CKSphere's offer to spy for them should be taken up. Under pressure from the Pentagon to acquire more intel on Soviet aircraft systems, in February 1978 CIA chiefs finally gave Hathaway the green light to respond to his overtures.

In March 1978, CKSphere handed over several handwritten notes to Hathaway containing details of Soviet radar development and a biography of himself. The CIA now realized that their new spy was one of the most important Soviet assets they had ever recruited. His name was Adolf Tolkachev, a senior technician specializing in military radar systems who worked at the USSR's main radar design and development facility, the NIIR (the Scientific Research Institute of Instrument Engineering) in Moscow, which was later renamed Phazotron.

Over the next seven years Tolkachev would provide the CIA with a wealth of top-secret information on Soviet radar and aircraft, which one US military officer referred to as a 'goldmine'. Although he was said to have been the most highly paid spy in the Agency's history, Tolkachev's motivation was thought to have been less financial than personal – he reportedly harboured a major grudge against the Soviet state, partly due to the fact that his wife's mother had been executed during the Great Terror.

At a secret meeting in December 1979, Tolkachev gave his CIA handler John Guilsher a notebook and photographs of classified documents on the Sapfir RP-23, as well as five circuit boards from the radar. He also told Guilsher that, as a direct result of Viktor Belenko's defection in the MiG-25 'Foxbat' three years earlier, Phazotron had been ordered to completely redesign and upgrade the MiG-25's Smerch-A radar, as its secrets had obviously been compromised.

Tolkachev went on to provide the Americans with extensive information, including design specifications, on the most advanced A.I. radar under development in the USSR, the N007 (NATO reporting name 'Flash Dance'). Intended to equip the MiG-31 'Foxhound' interceptor also then under development, work on the N007 – also known as the *Zaslon* (meaning 'Shield') – began in the early 1970s. It had a maximum detection range against fighters of 200km and could track low-flying targets such as cruise missiles at a range of up to 60km, while its onboard computer was capable of tracking up to ten targets and engage four of them simultaneously, using the MiG-31's R-33 (AA-9 'Amos') AAMs. The *Zaslon* was also highly resistant to electronic jamming. When it entered service in 1981, the combination of the MiG-31 and the *Zaslon* radar created what was considered at the time to be the world's most potent interceptor.

A CIA assessment from November 1983 stated: 'The Foxhound-A is the most advanced interceptor in the operational inventory [of the Soviet Air Force], incorporating a lookdown/shootdown capability with a track-while-scan and multiple target engagement capability. These new capabilities make the Foxhound a special threat to high speed, low altitude cruise missiles and aircraft attempting penetration of Soviet airspace.'[23]

At subsequent meetings with his handler, Tolkachev provided intel on the target recognition system of the new MiG-29 'Fulcrum' fighter and full details of an AWACS aircraft under development, the Beriev A-50 (called 'Mainstay' by NATO). Adapted from the Ilyushin Il-76 cargo plane and fitted with a powerful radar housed in a distinctive saucer-like radome atop the fuselage, which could track up to 50 targets simultaneously, when it entered service in 1985 the A-50 represented a major advance in the capabilities of Soviet air defences. However,

Tolkachev's revelation that the A-50's radar could not detect targets below 300 metres exposed a critical weakness in the system.

In 1980, Hathaway sent a memo to CIA director Stansfield Turner summarizing the intel CKSphere had thus far provided them with. This, wrote Hathaway, included 'The first documentation on the technical design characteristics of the new Soviet AWACS (it was CKSphere who first alerted us to the existence of this system and enabled us to locate it in overhead photography),' as well as 'extensive documentation on a new modification of the MiG-25, the first Soviet aircraft to be equipped with look-down/shoot-down radar,' and 'documentation on several new models of airborne missile systems and technical characteristics of other Soviet fighter and fighter/ bomber aircraft to be deployed between now and 1990.'[24]

By 1983, however, the KGB was beginning to suspect that there might be a spy at Phazotron. An initial investigation failed to identify the source of the leaks. But the KGB's suspicions were confirmed two years later by Aldrich Ames, a senior CIA officer who began selling information to the Soviets from April 1985. Ames revealed enough details of CKSphere for the KGB to identify him as Adolf Tolkachev.

On the night of Sunday, 9 June 1985, Tolkachev and his wife Natascha were returning to Moscow after spending the weekend at their dacha when their car was stopped at a police checkpoint. It was a trap. Tolkachev was grabbed by KGB officers posing as policemen, bundled into a van and taken to Lefortovo prison. Four days later, Tolkachev's latest CIA handler Paul Stombaugh was also arrested by the KGB as he arrived for a meeting with his agent in Moscow. Stombaugh's diplomatic status ensured his speedy release and expulsion from the USSR after being declared *persona non grata*. But Tolkachev enjoyed no such protection. Convicted of treason by a military tribunal, he was sentenced to death and executed in September 1986.

By then, however, Tolkachev's information had already thoroughly compromised the Soviet air defence network. Small wonder that within the CIA he was considered 'the billion dollar spy'.

Chapter Thirteen

Beyond the Cold War

By the mid-1980s, the Soviet Union was in crisis, bogged down in an unwinnable war in Afghanistan, the economy buckling under the strain of massive defence spending, and the populations of its satellite European states demanding democratic change. Recognising the need for urgent action, the new Soviet leader Mikhail Gorbachev began withdrawing troops from Afghanistan, slashing the defence budget, and introducing limited political reforms. Nothing, however, could now rescue the failing Soviet system. The end came quickly. The fall of the Berlin Wall in November 1989 and German reunification triggered a domino effect that saw the communist regimes throughout the Soviet bloc crumble one after the other in rapid succession. In February 1991 the Warsaw Pact was dissolved and in December the Soviet Union itself collapsed, to be replaced by the new Russian Federation. The Cold War was officially over.

Under the increasingly decrepit leadership of Boris Yeltsin, the years immediately following the dissolution of the USSR were marked by political and economic chaos in Russia. The collapse of the Soviet state inevitably had major repercussions for the intelligence services. Although the GRU survived – albeit with much reduced funding – the KGB was disbanded and its responsibilities split between two new organizations: the FSB (domestic intelligence) and the SVR (foreign intelligence).

The military was hit even harder. With budgets cut to the bone, ships were mothballed and aircraft of the newly constituted Russian Air Force were grounded, broken up for scrap or taken over by the former Soviet republics that had declared their independence, while soldiers, sailors and airmen went unpaid for months at a time. The country's massive aircraft industry was particularly badly affected during this tumultuous time. Military aircraft production, which had been running at about 800 per year in the 1980s, slowed to a trickle, with only those programmes deemed absolutely essential to the RuAF, such as the upgraded MiG-29M variant, allowed to continue. With few domestic military orders coming in, many OKBs had to diversify into the civil aviation sector or rely on export orders to survive.

With Russian military aircraft development having virtually ground to a halt, there was little demand for aviation S & T intelligence from the SVR and GRU, and operations in this area of activity greatly diminished.

Up close and personal

Meanwhile, the fall of the Iron Curtain permitted Western observers to gain access to Soviet bloc aircraft. In 1991, following German reunification, aviation photo-journalist Peter R. Foster visited the former GDR on a planespotting trip to take photographs of aircraft at the Soviet Air Force bases in the east of Germany as they were preparing to depart the country. Just a few months before, these aircraft were strictly off limits to Westerners and those found snooping around East German bases were liable to be shot, or at the very least arrested. But with the political climate completely changed, Foster found that he could move around relatively freely and take his photographs of Su-24s, MiG-23s and MiG-25s without much interference from the Soviet guards. 'Not only could we now reach the former Soviet satellite states that had quickly disengaged themselves from Moscow's influence, but also get a taste of Soviet aviation without too much fear of repercussions,' Foster explained.[1]

American technicians also gained access to Russia's most formidable strategic bomber, the Tupolev Tu-160 'Blackjack'. The bulk of the Tu-160 fleet was stationed at the Priluki air base in Ukraine at the time of the break-up of the USSR. As the air force of the newly independent Ukraine had neither the resources to maintain nor an operational requirement for these aircraft, in 1998 the Kyiv government contracted the US Raytheon corporation to dismantle some of the bombers. Protracted negotiations between Kyiv and Moscow saved eight of the Tu-160s from the breaker's torch, these being handed back to Russia in return for debts owed for natural gas supplies being cancelled.

Pilots of the West's air forces, too, benefitted from the sudden availability of Soviet warplanes. During the Bosnian war of the 1990s, the UN imposed a no-fly zone over Bosnia, enforced by NATO aircraft (Operation Deny Flight), and the prospect of aerial clashes between the Serbian JRV (Federal Yugoslav Air Force) and NATO became very real. The most advanced fighter in the Serbs' inventory was the MiG-29, and NATO airmen took the opportunity to hone their skills in mock dogfights against former LSK MiG-29s, which had been absorbed by the Luftwaffe, to prepare them for the possibility of air combat with the Serbs.

One pilot who participated in these exercises, held at the NATO air base at Decimomannu in Sardinia, was Lieutenant Nick Richardson, a Sea Harrier FRS.1 pilot serving with the Royal Navy's No 801 Squadron aboard HMS *Ark Royal*. Richardson was impressed by the performance of the MiG, and in mock dogfights between the Fulcrum and his Sea Harrier, the Russian fighter invariably emerged victorious.[2]

Relations between Moscow and the West during the early years of the Yeltsin presidency had become so friendly that USAF pilots were even invited to Russia to test fly the MiG-29 and other Russian types. In 1993 a team of five US test pilots arrived at the Gromov Flight Research Institute to put the MiG-29, Sukhoi Su-27 and MiG-25 through their paces. For these Cold War veterans, who had only seen these aircraft in photographs or from a distance, finding themselves in the cockpit flying the aircraft of their erstwhile enemy was, according to one of the US airmen, an 'unbelievable' experience.[3] A few months later the Americans returned the favour by inviting a group of Russian test pilots to Edwards AFB in California to fly the F-15 and F-16.

But this new spirit of cooperation was not to last. Within a few years Cold War-style tensions were resurfacing, principally over NATO enlargement in Eastern Europe – which Moscow still regarded as its rightful sphere of influence – and the NATO air campaign over Kosovo in 1999.

The Putin years

These tensions grew more acute when the hardliner Vladimir Putin, a former KGB officer and head of the FSB, became president of the Russian Federation in 2000. A priority for Putin was rebuilding Russia's moribund armed forces and aircraft industry. After years of rundown, the Russian Air Force once again became a credible force. By 2007 Putin felt sufficiently confident in his rejuvenated military to challenge NATO's global hegemony. This confidence manifested itself in the deployment of Russian airpower to intervene in the Syrian Civil War and resuming the old Cold War practice of despatching long-range bombers to probe the airspace of NATO member states.

A key element of Putin's modernization of the Russian aircraft industry, which had fallen very far behind its Western counterparts during the years of neglect, was the reactivation of S & T intelligence-gathering. This began even before he became president. On the night of 27 March 1999, during the early stages of Operation Allied Force – the NATO air campaign against Serbia to halt ethnic cleansing in the contested province of Kosovo – an F-117 Nighthawk (callsign 'Vega 31') of the 49th Fighter Wing, based at Aviano in northern

Italy, was shot down by an S-125 (SA-3 'Goa') SAM near the village of Budanovci, around 50km from Belgrade. The pilot, Lieutenant Colonel Dale Zelko, ejected and was swiftly picked up by a USAF search and rescue helicopter. But the wreckage of one of the world's most advanced warplanes was now in Serb hands.

The first combat loss of an F-117 was an embarrassment to the USAF and a major propaganda coup for the Serbs.[4] Some of the wreckage was proudly put on display at the Museum of Aviation in Belgrade. It has also been widely reported that within days of the shootdown, Russian specialists visited the crash-site and other pieces of wreckage were handed over to them by the Serbs to take back to Russia for analysis.

The knowledge gained from the F-117 wreckage by Russian aerospace engineers almost certainly influenced the development of the country's first true stealth aircraft, Sukhoi's Su-57 multirole fighter (NATO reporting name 'Felon').

Work began on this project in 1999 when the Russian Ministry of Defence issued a specification for a fifth-generation fighter, intended to eventually replace both the MiG-29 and Su-27. Sukhoi's submission, the T-50, was selected over MiG's rival entry, the E-721, in 2002 and the prototype made its maiden flight from the Dzyomgi air base in Russia's Far East on 29 January 2010. Broadly comparable to the Lockheed Martin F-22 Raptor, low-rate production of what was now designated the Su-57 began in 2019, with the first examples entering service with the Russian Air Force a year later.

The Su-57 has seen some limited operational service in the Russo-Ukrainian War, but with only a small fleet available – thought to number only a few dozen – and the high unit cost, the Kremlin was believed to be reluctant to risk the Su-57 in the more high-threat areas of the conflict zone. Despite this, two Su-57s were claimed damaged by Ukraine in a drone attack on their base at Akhtubinsk in the Astrakhan region in June 2024.

Both the SVR and GRU have been stepping up their efforts to steal information on Western aviation technology in recent years. In June 2021, a 29-year-old Russian researcher at the Bavarian University of Augsburg, identified in court only as Ilnur 'N', was arrested on suspicion of passing information regarding European aerospace projects to the SVR, including on the Ariane 6 space rocket programme. He was found guilty in April 2022 in a court in Munich and given a one-year suspended sentence. It was reported that he had been paid just €2,500 by the SVR.

One aircraft of particular interest to Russia is the Lockheed Martin F-35 Lightning II Joint Strike Fighter. The West's most advanced fighter, the F-35 is the latest generation stealth aircraft, designed to operate both from land bases and aircraft carriers (for which a STOVL version, the F-35B, was developed). Entering service in 2015, the F-35 is operated by the US Navy and Marine Corps, the RAF and Fleet Air Arm, and many other NATO member states.

On 17 November 2021, an F-35B (serial ZM152) of the RAF's No 617 Squadron crashed into the Mediterranean after taking off from the aircraft carrier HMS *Queen Elizabeth II*, due to an engine intake plug being inadvertently left in. The pilot safely ejected. In an echo of the crash of an F-14 Tomcat off Scotland in 1976 (see Chapter Eleven), fears that the Russians may try to salvage the wreckage prompted a major joint UK-US operation to recover the wreck, using a remotely operated vehicle to locate ZM152. 'It was later discovered intact, inverted on the seabed, at a depth of 2,000m, with a few minor parts such as the ejection seat detached but close to the airframe. A salvage operation recovered the aircraft, and all the detached items, and then transported them back to the UK,' stated the official UK Service Inquiry into the loss of ZM152. The cost of the salvage operation was put at £2.6 million.[5]

As well as the more traditional methods of espionage, the twenty-first century has witnessed the rapid growth of hacking as a means of stealing data on the latest Western aviation technologies, with the Russian and Chinese intelligence services at the forefront of this information collection effort. In 2024 the NSA warned of this threat: 'Since at least 2021, the SVR ... have consistently targeted US, European and global entities in the defense, technology and finance sectors.' The chief targets of the SVR hackers identified by the NSA were government departments, technology companies, think tanks and defence contractors. 'These groups are targeted for the purpose of collecting foreign intelligence and technical data,' the NSA confirmed.[6]

The Russo-Ukrainian War, which began in February 2022, has also presented opportunities for both sides to capture and examine the latest aviation technology. After gaining approval from the US government for their re-export following months of lobbying by Kyiv, Denmark, Norway, Belgium and the Netherlands began providing some of their Lockheed Martin F-16Ms to Ukraine, the first batch – ex-Royal Danish Air Force examples – arriving in July 2024. Five months later the Ukrainian internal intelligence service, the SBU, broke up an alleged Russian spy ring in the country which was gathering intelligence on the donated F-16s and their weapons systems.

One of the most successful weapons systems donated by the West to the Ukrainians was the Storm Shadow. Jointly developed by Britain and France in the 1990s, the Storm Shadow is a long-range cruise missile, designed for use on tactical strike aircraft like the Tornado, Mirage 2000, Rafale and Typhoon. Deliveries of Storm Shadow missiles from the UK to the Ukrainian Air Force began in early 2023, and it has been used with considerable success, fired from specially adapted UAF Su-24 'Fencer' bombers. Among the targets hit were the strategically important Chonhar road bridge in June 2023 and the Russian landing ship *Minsk* in September 2023. In the summer of 2023, a Storm Shadow was retrieved almost intact by Russian forces in a field near Zaporizhzhia. Despite Ukrainian attempts to interfere with the recovery operation, the missile was evacuated safely to Moscow, where, according to the Russian TASS news agency, it would be analysed by technical specialists to aid the development of countermeasures.[7]

The Ukrainians have enjoyed their own successes in capturing the enemy's aviation hardware. In an exploit reminiscent of the classic missions of the Cold War, such as Operations Moolah and Diamond, Ukrainian intelligence successfully orchestrated the defection of a Russian pilot with his Mil Mi-8 helicopter. The ambitious operation began in early 2023 when Captain Maxim Kuzminov, a pilot in Russian Army Aviation's 319th Separate Helicopter Regiment, disillusioned with the war, secretly made contact with Ukraine's HUR (the Main Directorate of Intelligence), offering to defect and bring with him an Mi-8. Although the Mi-8 was an ageing design, having first entered Soviet service in the 1960s, Kuzminov's helicopter was the latest version, built in 2016, and so its capture would be a significant coup for Kyiv.

After a deal was negotiated with Kuzminov, which included political asylum for himself and his family and a reward of $500,000, a plan was worked out with the HUR over the next few months, known as Operation *Synytsia*. Once his family was safely taken out of Russia, the plan was put into action. On the afternoon of 9 August 2023, Kuzminov took off from an air base at Kursk, carrying a cargo of spare parts for Russian Su-27, Su-30 and Su-35 fighters. He soon deviated from his flight path and turned off his radio. The other two members of his crew realized he was trying to defect but as Kuzminov was the only pilot amongst them could do nothing to stop him. Flying at a height of just 10 metres, he came under fire from soldiers on the ground. 'When crossing the border they started firing at me,' recalled Kuzminov, 'I can't say for sure who led it, but I assume it was the

Russian side. I was wounded in the leg by small arms fire.'[8] He flew on for another 20km and put the Mi-8 down at a pre-arranged location near Poltava. The Ukrainians stated that the other two crewmen were shot while trying to escape back to their own lines once the helicopter had landed. 'When they realized where they had landed, they attempted to flee. Unfortunately, they were neutralized; we wanted to take them alive, but we have what we have,' said Kyrylo Budanov, chief of the HUR.[9]

Even more important than the helicopter itself was the cargo of aircraft parts it was carrying, which Andriy Yusov, a spokesman for the HUR, revealed had given them 'invaluable information' about the Russian Air Force's aircraft.[10]

Russian aerial weapons have also fallen into Ukrainian hands. In February 2023 they recovered remnants of a Vympel R-37M AAM from the battlefield and reportedly handed this over to the British for study.[11] The R-37M (known to NATO as the AA-13 'Axehead') is an ultra-long-range air-to-air missile, whose capabilities are similar to that of the US AIM-54 Phoenix. Development of the missile began in the 1980s, for use with the MiG-31 interceptor, but was suspended after the break-up of the USSR due to budget cuts. Work on the R-37M resumed in 2006 and the missile eventually entered service in 2019. With a maximum range of 200km and capable of hitting targets at altitudes up to 82,000ft, the R-37M is one of the world's most potent AAMs and much feared by the Ukrainian Air Force. As well as the MiG-31, the R-37M is also used on the Su-35 and Su-57.

Other weapons retrieved from the battlefield by the Ukrainians reportedly included an example of the Kh-101, a stand-off, air-launched cruise missile known to NATO as the 'Kodiak'. Launched from strategic bombers like the Tu-95MSM and Tu-160M, the Kh-101 has a 400kg warhead and a maximum range of 5,500km. The missile was recovered, almost intact, by Ukrainian forces in January 2023.

Rise of the dragon

Lacking the Soviet Union's industrial infrastructure, throughout the Cold War the People's Republic of China was the junior partner in the often fraught alliance between the two communist powers. Since the collapse of the Soviet Union and its replacement by the Russian Federation, however, that situation has changed, with China overtaking its sometime Cold War ally economically, industrially and militarily. From the 1990s onwards, the scale of Chinese industrial espionage in the West has also eclipsed that of the Russian Federation. The FBI has warned that '… the Chinese government is seeking to become the world's greatest superpower through predatory lending and business practices, systematic theft of intellectual property and brazen cyber intrusions.'[12]

As with Russia, the West's aviation industries are high on the target list of the Chinese intelligence service, the MSS, with US stealth technology being a priority target.

According to some reports, besides the Russians, the Chinese also obtained parts of Lieutenant Colonel Zelko's F-117 that was brought down over Serbia during Operation Allied Force in 1999. In 2011, Admiral Davor Domazet-Lošo, a former chief of Croatian military intelligence, insisted that 'our intelligence reports told of Chinese agents criss-crossing the region where the F-117 disintegrated, buying up parts of the plane from local farmers. We believe the Chinese used those materials to gain an insight into secret stealth technologies … and to reverse-engineer them.'[13]

Other reports suggest that the Chinese secretly negotiated a deal with the Serbs for parts of the wreckage in exchange for facilitating Yugoslav military communications through their embassy in Belgrade during the war. This has led to speculation that the attack on the embassy on 7 May 1999 by a USAF B-2 Spirit stealth bomber, which killed three Chinese journalists, may not have been an accident resulting from the use of outdated maps (as the official US report into the incident concluded) but rather an intentional act to halt these communications. An alternate, more outlandish, conspiracy theory is that the attack was an attempt to destroy pieces of the F-117 wreckage being stored in the basement of the embassy before they could be shipped back to China. According to this conspiracy, one of the laser-guided J-DAMs (Joint Direct Attack Munitions) dropped by the B-2 penetrated the basement but failed to detonate.[14]

Chinese intelligence gained further secrets of US stealth technology thanks to a spy they recruited at the heart of the B-2 programme. Noshir Gowadia was born in Mumbai, India in 1944 but later emigrated to America and became a US citizen. In 1968 he joined Northrop Grumman and became an important figure in the development of the company's B-2, designing the propulsion system that allowed the aircraft to evade infrared detection. After leaving the company in 1986, he set up his own aerospace consultancy business. From 2003 to 2005 he made six trips to China, arousing the suspicions of the FBI, which launched a joint investigation with the USAF's Office of Special Investigations. A search of his luxury home on the Hawaiian island of Maui uncovered 40 boxes of classified documents, many relating to stealth. Arrested in October 2005, he was found guilty at the US District Court of Hawaii in August 2010 of 14 of the 17 charges made against him, including 'unlawfully exporting classified information about the B-2' and 'agreeing to design, and later designing, a low observable cruise missile exhaust system nozzle capable of rendering the missile less susceptible to detection and interception', for which he was paid $110,000 by the Chinese.

'Mr Gowadia went beyond disclosing information to China; he performed defence work in that nation with the purpose of assisting them in their stealth weapons design programs,' said US Attorney for the District of Hawaii Florence Nakakuni after his conviction. He was sentenced to 32 years imprisonment.[15]

Chinese intelligence has also been hard at work procuring the secrets of the latest generation of US stealth fighters, the Lockheed Martin F-22 Raptor and F-35. In April 2018 Xu Yanjun, a senior officer in the MSS, was arrested in Belgium and extradited to the US on charges of attempting to recruit an engineer at the aerospace giant GE Aviation (which supplies many of the electronic systems used in the F-35). Yanjun was convicted in 2021 and received a 20-year prison sentence.

Beijing's campaign of cyber espionage, meanwhile, has been conducted on a truly industrial scale, dwarfing that of Russia. Frank Kendall, the Under Secretary of Defense and Acquisitions at the Pentagon, confirmed to a Senate hearing in 2013 that Chinese hackers were gathering intel on the F-35 programme. 'A lot of that is being stolen right now and it's a major problem for us,' Kendall admitted. 'What [cyber espionage] does is reduce the costs and lead time of our adversaries to doing their own designs, so it gives away a substantial advantage.'[16] The following year the FBI's Cyber Division smashed a major Chinese hacking operation run by the PLA targeting the US aviation industry, with the assistance of a Chinese national resident in Canada, Su Bin (also known as Stephen Su), who owned an aircraft parts supply company. The FBI accused Su of helping the PLA hack more than 630,000 files from US aerospace companies over a six-year period. Amongst the data gathered by Su and his associates were technical specifications and blueprints of the Boeing C-17 cargo plane and the F-22 and F-35 fighters. In an email he sent to one of his co-conspirators, Su boasted that the information on the latter would allow China to 'catch up rapidly with U.S. levels … [and] stand easily on the giant's shoulders.'[17] At his trial in 2016, he pleaded guilty to the charge of conspiring to gain unauthorized access to protected computer networks in the US to obtain military information and received a 46-month prison term.

Chinese targets of industrial espionage have not just been confined to the US. In 2019 Yevgeny Livadny, chief of intellectual property at the Russian defence conglomerate Rostec, complained that China had illegally copied Russian aircraft engines, air defence systems and Sukhoi fighters, including the Su-33 – which the Chinese reverse-engineered to produce the Shenyang J-15 carrier-borne fighter – and Su-57.

Chinese interest in stealth aircraft has also extended to the field of rotary-wing aviation, and access to top secret American stealth helicopter technology is believed to have come China's way thanks to the most high-profile special forces mission of recent times. In the early hours of 2 May 2011, two dozen US Navy SEALs raided the compound of Osama bin Laden in Abbottabad, Pakistan, killing the al-Qaeda terror chief, in what was called Operation Neptune Spear. The SEAL team was flown to their target in two heavily modified UH-60 Black Hawk helicopters of the 160th Special Operations Aviation Regiment, which incorporated advanced stealth design features. During the mission, one of the Black Hawks crashed within the grounds of the compound after its tail rotor hit a wall. To protect the secrets of the aircraft – whose existence had not been officially disclosed to the public – members of the SEAL team destroyed it with explosive charges. Unfortunately, parts of the Black Hawk survived, including the tail section.

Pakistani soldiers swiftly recovered these parts and moved them to a secret location. After negotiations with the US State Department, the Pakistani government returned the wreckage to the Americans two weeks later. However, newspaper reports claimed that the Pakistanis – who had not been consulted by the Americans in advance of the bin Laden raid and were furious at the unauthorized military action on their territory – agreed to a Chinese request to secretly examine the helicopter parts before they were handed back. These claims were strenuously denied by the Pakistani government, which described them as 'baseless and speculative'.[18] Lending credence to the rumours that the Chinese gained access to the stealthy UH-60 from the Pakistanis was the news in 2015 that they were working on a stealth version of their Harbin Z-20, a medium-lift helicopter, which itself is rumoured to be a reverse-engineered copy of the Sikorsky S-70, the export version of the Black Hawk (of which China bought 24 in the mid-1980s, when US-Chinese relations were somewhat warmer). A concept model of the stealth version of the Z-20 was publicly unveiled in 2021.

Beijing's intensive intelligence effort to acquire the secrets of US and Russian stealth technology since the 1990s, through the examination of downed aircraft, cyber espionage and the recruitment of spies in the aerospace sector, has almost certainly made a major contribution to the design and development of China's stealth aircraft projects.

The first of these was the Chengdu J-20 'Mighty Dragon', a multi-role fighter given the reporting name 'Fagin' by NATO, which made its maiden flight on 11 January 2011 and began entering PLAAF service six years later. This was followed by the Shenyang J-35, a stealth fighter developed in both land and carrier-based variants. The J-35's striking resemblance to its American near-namesake, Lockheed Martin's F-35, prompted John Bolton, then US national security adviser to President Trump, to remark in 2019: 'It looks a lot like the F-35 – because it is the F-35. They just stole it.'[19] Beijing also officially confirmed in 2016 that it had a strategic stealth bomber under development, the Xian H-20. Few hard facts about this project have so far emerged, although it is likely the H-20 will be comparable in capability to the B-2 Spirit. The H-20 is expected to enter service with the PLAAF during the 2030s.

Cases like that of Xu Yanjun and Su Bin led the chiefs of the intelligence agencies of the 'Five Eyes' alliance to issue a joint warning at a security conference held in California in October 2023 of the threat posed to the West by Chinese industrial espionage. 'We have seen a sustained [Chinese] campaign on a pretty epic scale,' said MI5 Director-General Ken McCallum. FBI Director Christopher Wray added that, 'China has made economic espionage and stealing others' work and ideas a central component of its national strategy and that espionage is at the expense of innovators in all five of our countries. That threat has only gotten more dangerous and more insidious in recent years.'[20]

Conclusion

The battle for aviation technology was a major element of the intelligence war fought covertly with ceaseless vigour by the spy agencies of the NATO and Warsaw Pact member states – and their allies – throughout the Cold War. For the Soviets, this battle began long before the Cold War officially began, as soon as the USSR itself came into being, in fact.

This is hardly surprising. Stalin regarded a strong indigenous aviation industry as central to his vision of a modern, industrialized Soviet Union, and saw S & T intelligence as a way of expediting this goal.

In the golden age of Soviet espionage, which spanned the 1930s to the early 1950s, when communism enjoyed considerable allure in the West, the NKVD and GRU were able to recruit many highly placed individuals in the North American and West European aviation industries and air forces willing to spy for the USSR. After Stalin's death, these agencies found it much more difficult to recruit assets motivated by ideological conviction, forcing them to rely increasingly on less dependable mercenary traitors, who traded their countries' secrets only for financial reward.

On the other side of the Iron Curtain, the situation was very different. The authoritarian nature of the USSR and its satellite states led to a steady stream of disaffected individuals occupying senior positions in the intelligence services and industry who enthusiastically passed on intelligence of the highest value to the Americans, British and French. Among the most important of these were the Directorate T officer Vladimir Vetrov and the radar designer

Adolf Tolkachev. Disillusionment also infected the air forces of the Soviet bloc and their communist allies, with numerous pilot defectors fleeing to the West in their prized aircraft, the most famous of whom was Lieutenant Viktor Belenko.

Nikita Khrushchev, Soviet leader from 1953 until 1964, was confident the USSR would eventually overtake its main rival, the USA, in the technology race. With the Soviets' successful launch of the world's first satellite, Sputnik-1, in October 1957, and putting the first man into space four years later, this confidence seemed well-founded. Khrushchev, according to a CIA report, was 'sincerely hopeful that Soviet technological prowess – revealed most dramatically by the advent of Sputnik in 1957 – would enable the Soviets to surpass the West by 1980.'[1]

Khrushchev's optimism proved misplaced, however. Although Soviet espionage played a crucial role in helping the USSR narrow the technology gap that always existed between East and West, it could never fully close it. Indeed, far from surpassing American technological achievements, by 1980 that gap was widening, despite the vast resources being poured into the defence industry by the Soviet leadership.

The problem did not lie with the Soviet Union's intelligence services – by the 1980s Directorate T and the GRU were gathering masses of high-grade intel on the West's latest military and aviation technologies. The blame instead lay chiefly with Soviet industry and the inflexible communist system, which were incapable of fully exploiting that intel to produce hardware comparable in quality to that of the United States and Western Europe. This problem only became more acute as technology became more advanced. A case in point is stealth. Although Moscow received considerable intelligence on US stealth technology during the 1980s, the Soviets were unable to produce their own stealth warplane before the end of the Cold War.

When Soviet aircraft fell into Western hands, the technical specialists who examined them were more often than not left unimpressed, judging many of their aircraft to be generally inferior technically to their American and West European equivalents, and quite crude in their construction. 'Nevertheless,' said Duane Clarridge, a CIA officer involved in S & T collection work in the 1970s, 'they had some sophisticated design features that made them of interest not only to the Air Force but to US industry as well.'[2]

The Soviet track record in reverse-engineering Western aircraft, engines and weapons systems was also decidedly uneven. In the early years of the Cold War there were several notable successes, such as the Tupolev Tu-4 'Bull' (a B-29 clone), the Klimov VK-1 gas-turbine engine (a near-exact copy of the Rolls-Royce Nene) and the K-13 AAM (based on the AIM-9B Sidewinder). As technology became more sophisticated, however, Soviet engineers found it increasingly difficult to reverse-engineer Western aircraft and other hardware. Examples of this were the Beriev S-13 (based on the wreckage of Francis Gary Powers' Lockheed U-2 spy plane, shot down near Sverdlovsk in 1960) and the Tupolev '*Voron*', a failed attempt to copy another reconnaissance platform from the Lockheed works that fell into Soviet hands, the D-21 drone. As a CIA intelligence report from 1986 observed: 'Although the Soviets reap substantial benefits from imported technology, they frequently have problems assimilating it into production. These difficulties are sometimes greater when the technology is illegally acquired, because in those cases Soviet engineers usually cannot benefit from foreign training and technical assistance. Modern critical technologies and equipment are generally more difficult to transfer – and much more difficult to duplicate by reverse engineering – than those that contributed to early Soviet industrial development.'[3]

The fundamental weakness of Soviet reliance on foreign technology had been pointed out by Vladimir Vetrov, the Directorate T officer who spied for the French, to his handler. 'It's just like a bad student copying from his neighbour,' he remarked. 'When he can no longer copy, whatever the reason, he has no alternative solution.'[4] It was also true that by the time the Soviets brought their own copies of Western designs into service, many were already obsolescent – a flaw of the Soviet reverse-engineering policy that Andrei Tupolev himself had warned of back in the 1930s.[5]

The most important question, of course, is what effect did all this S & T intel both sides went to such lengths to obtain have on the frontline? Though, thankfully, direct military confrontation between NATO and the Warsaw Pact never materialized, there were numerous proxy conflicts during the Cold War that pitted the machines of East and West against each other.

In Korea, the introduction of the MiG-15 allowed the communists to seriously challenge UN air supremacy over the peninsula, prompting the Americans to rush their most advanced fighter, the F-86 Sabre, to the warzone. The great efforts made by the Americans and the British to capture the wreckage, and subsequently a flyable example, of the MiG-15 attests to the fear this fighter generated in Western military minds. In Vietnam, Soviet exploitation of equipment recovered from downed US aircraft, especially ECM equipment, played an important role in helping the North Vietnamese counter the increasingly technically sophisticated air offensive being waged by the United States, in what amounted to a South-East Asian version of the 'wizard war'.[6] On the American side, knowledge gained of the MiG-17 and MiG-21 through the Have Drill and Have Doughnut programmes contributed to a significant improvement in the performance of US fighter pilots in the latter stages of the air war over North Vietnam.

But it was in the Middle East where aviation S & T had perhaps the greatest impact. Mossad's 1966 intelligence coup, orchestrating the defection of an Iraqi pilot in his MiG-21, played a key role in Israel's stunning victory over her Arab neighbours in the 1967 Six Day War. 'That MiG had an important part in the victory of the Israeli Air Force over the Arab air forces, and in particular in the destruction of the Egyptian Air Force in a few hours,' asserted the then Mossad chief, Meir Amit, who supervised the defection operation.[7] It was a similar story in the Yom Kippur War six years later, when IAF F-4 Phantoms dominated in air combat against the MiG-21s of the Arab coalition. Analysis of the defector Lieutenant Belenko's MiG-25 and intelligence gathered on the MiG-29, especially its radar (courtesy of the CIA source Adolf Tolkachev), also helped US airmen sweep the mainly Soviet-equipped Iraqi Air Force from the skies in the opening days of 1991's Operation Desert Storm.

Far from diminishing, since the end of the Cold War the importance of S & T has grown as relations between the NATO states, Russia and China have steadily deteriorated. S & T collection in the West – especially in the field of stealth technology – has been a key element of China's massive military expansion and modernization in the twenty-first century, as Beijing seeks to challenge the USA's position as the world's sole superpower. Moscow's ambitions to regain its lost superpower status, meanwhile, has led Russia's intelligence services to resume their campaign of industrial larceny.

With a new Cold War having taken hold between East and West, the battle for the secrets of each side's aviation technology looks set not only to continue, but to intensify.

Endnotes

Introduction

1. Johnson, *Secret Agencies*, p. 176.
2. Weiss, *Duping the Soviets – the Farewell Dossier.*
3. Ibid.
4. Andrew and Mitrokhin, *The Mitrokhin Archive*, p. 723.
5. www.cia.gov/readingroom ('The Soviet Weapons Industry: An Overview', September 1986).
6. www.cia.gov/readingroom ('Soviet Acquisition of Western Technology', April 1982).
7. Ibid.

Chapter One

1. www.people.uncw.edu.
2. Dancey, *Soviet Aircraft Industry*, p. 11.
3. Bailes, *Technology and Science Under Lenin and Stalin*, p. 399.
4. National Archives, Kew: KV 2/643.
5. National Archives, Kew: KV 2/647.
6. National Archives, Kew: KV 2/770.
7. National Archives, Kew: KV 2/1024 and KV 2/1196.
8. The National Minority Movement was founded by the CPGB in 1924, with the aim of infiltrating industry and the trade unions.
9. National Archives, Kew: KV 2/1024.
10. National Archives, Kew: KV 2/1196.
11. Ibid.
12. National Archives, Kew: KV 2/2029 and KV 2/2031.
13. National Archives, Kew: KV 2/996.
14. National Archives, Kew: KV 2/2199.
15. Ibid.

16. National Archives, Kew: KV 2/996.

17. *FlyPast*, February 2014 issue, No. 391.

18. Haynes *et al*, *Spies: Rise and Fall of the KGB in America*, p. 331.

19. Ibid., p. 370.

20. Lokhova, *The Spy Who Changed History*, p. 239.

21. www.cia.gov/readingroom ('Venona: Soviet Espionage and the American Response 1939 – 1957').

22. Op. cit., *The Spy Who Changed History*, p. 367.

Chapter Two

1. BBC radio broadcast, 22 June 1941.

2. Drabkin, *Barbarossa and the Retreat to Moscow*, pp. 11, 129.

3. National Archives, Kew: CAB 79/15/24.

4. National Archives, Kew: CAB 65/33/32.

5. Deane, *The Strange Alliance*, p. 326.

6. Brown, *Wings on my Sleeve*, p. 112 and pp. 121-22.

7. *Luftwaffe Secret Bombers of the Third Reich*, p. 120

8. National Archives, Kew: CAB 80/35.

9. National Archives, Kew: CAB 79/19/16.

10. West, *The A to Z of British Intelligence*, p. 34.

11. National Archives, Kew: KV 2/1596 and KV 2/4600.

12. Hastings, *The Secret War*, p. 349.

13. The *Tirpitz* was finally sunk on a subsequent raid by RAF Lancasters two months later.

14. National Archives, Kew: CAB 79/82/10.

15. Lokhova, *The Spy Who Changed History*, pp. 358-59.

16. https://media.defense.gov (Venona signal decrypt, 31 October 1943).

17. Op. cit., *The Spy Who Changed History*, p. 392.

18. In 1945 Mironov was executed after trying to expose the truth that the Soviets rather than the Germans were responsible for the infamous 1940 Katyn massacre in Poland.

19. Andrew and Mitrokhin, *The Mitrokhin Archive*, p. 173.

Chapter Three

1. Andrew and Mirokhin, *The Mitrokhin Archive*, p. 150

2. Ibid. p. 168.

3. Ibid. p. 154.
4. A labour camp run by the NKVD specifically for engineers and scientists, where the prisoners were compelled to continue their work.
5. West, *Soviet and Nazi Defectors of WW2*, p. 24.
6. Reshetnikov, *Bomber Pilot on the Eastern Front*, p. 171.
7. Duffy and Kanalov, *Tupolev: The Man and his Aircraft*, p. 12.
8. In 1991 the LII was renamed the Gromov Flight Research Institute in his honour.
9. Gordon and Rigmant, *Tupolev Tu-4 – Soviet Superfortress*, p. 36.

Chapter Four

1. National Archives, Kew: CAB 79/23/35.
2. Lokhova, *The Spy Who Changed History*, p. 364.
3. *'Soviet Espionage Activities in Connection with Jet Propulsion'* (Committee of Un-American Activities report, 1949).
4. Feklisov and Kostine, *The Man Behind the Rosenbergs*, p. 138.
5. National Archives, Kew: HW 15/22/70.
6. Holder, *Lost Fighters*, p. 14.
7. National Archives, Kew: HW 15/25/22.
8. Haynes et al, *Spies: The Rise and Fall of the KGB in America*, p. 340.
9. Andrew and Mitrokhin, *The Mitrokhin Archive*, pp. 164-65.
10. *Luftwaffe Secret Bombers of the Third Reich*, p. 120.
11. *Aeroplane Monthly*, June 2017, issue no. 530.
12. Ibid.
13. National Archives, Kew: FO 371/94874.
14. National Archives, Kew: AIR 22/93/1.
15. Ibid.
16. *'Patterns of Communist Espionage'* (Committee of Un-American Activities report, 1959).
17. Hansard, 12 March 1951.
18. National Archives, Kew: FO 371/94874.
19. Hansard, 27 April 1953.

Chapter Five

1. Lieutenant Brown's MiG-15 'kill' remains disputed, Soviet accounts claiming that the MiG escaped and landed safely back at its base at Mukden in China.

2. Caygill, *Meteor From the Cockpit*, pp. 44-5.

3. *MiG Red Star Fighters*, p. 29.

4. Harden, *King of Spies*, p. 4.

5. www.valor.militarytimes.com.

6. Op. cit., *King of Spies*, p. 102.

7. https://nsarchive.gwu.edu ('US Intelligence Summary,' 24 June 1951).

8. Op. Cit., *King of Spies*, p. 112.

9. National Archives, Kew: ADM 1/22578.

10. https://nsarchive.gwu.edu ('Far East Air Force Intelligence Roundup,' 12 August 1951).

11. Op. cit., ADM 1/22578.

12. https://nsarchive.gwu.edu ('Intelligence from aircraft acquired in Korea,' 22 July 1953).

13. Ibid. ('Early History of the CIA's Sovmat Staff,' undated).

14. Ashcroft, *Heroes of the Skies*, p. 184.

15. Jeka was killed in April 1958 when his Douglas A-26 Invader crashed on the Indonesian island of Sulawesi, whilst on a covert CIA mission to support anti-communist rebels.

16. www.cia.gov/readingroom (CIA Intelligence Bulletin, 6 March 1953).

17. National Archives, Kew: CAB 128/26/18.

18. www.cia.gov/readingroom (CIA Intelligence Bulletin, 11 March 1953).

19. www.archive.gov ('Zdzislaw (Jerzy) Jazwinski memorandum', undated).

20. www.governmentattic.org ('ATIC technical report TR-AE-63 MiG-15 (VK-1),' 26 April 1955).

Chapter Six

1. National Archives, Kew: AIR 21/13016.

2. Ibid.

3. National Archives, Kew: AIR 27/2388/3.

4. Various Air Ministry documents relating to crash of Meteor T.7 (WA685) and Soviet detention of pilot, obtained via Freedom of Information request (reference: FOI 2023/10894).

5. Ibid.

6. National Archives, Kew: AIR 27/2854/49.

7. *Lodi News-Sentinel*, 27 May 1967.

8. National Archives, Kew: DEFE 44/482.

9. www.cia.gov/readingroom ('Evolution of the Mikoyan design bureau since 1970s and MiG-25,' 24 January 1990).
10. *MiG: Red Star Fighters.*
11. Ibid.
12. *New York Times*, 20 September 1976.
13. history.state.gov/historicaldocuments.
14. Op. cit., 'Evolution of the Mikoyan design bureau since 1970s and MiG-25'.
15. history.state.gov/historicaldocuments.
16. Viktor Belenko died of natural causes aged 76 in 2023.
17. *The Philadelphia Enquirer*, 11 May 1981.
18. https://nsarchive.gwu.edu ('F-6/MiG-19 Exploitation in Taiwan', 16 March 1978).
19. *Los Angeles Times*, 12 October 1989.
20. https://en.wikipedia.org/wiki/Aleksandr_Zuyev_(pilot).
21. Some newspaper reports from the time suggest that US experts were secretly granted brief access to Captain Zuyev's MiG-29 before it was handed over to the Soviets.
22. *Los Angeles Times*, 21 May 1989.
23. Zuyev's book *Fulcrum: A Top Gun Pilot's Escape from the Soviet Empire* was published in 1992.

Chapter Seven

1. *Patterns of Communist Espionage* (Committee of Un-American Activities report, 1959).
2. Andrew and Mitrokhin, *The Mitrokhin Archive*, p. 219.
3. US Congressional Record, Proceedings and Debates of the 92nd Congress Second Session, 16 May 1972.
4. Gordon and Rigmant, *OKB Tupolev - A History of the Design Bureau and its Aircraft*, p. 235-236.
5. In his memoirs, senior Lockheed engineer Ben Rich, who had worked on the D-21 project, revealed that in February 1986 a CIA officer handed him a panel from the engine mount of the D-21 that had crashed in Mongolia and told him that he'd received it from a KGB agent.
6. The BAC TSR 2 was ultimately cancelled in April 1965.
7. Andrew, *Defence of the Realm*, p. 583.
8. *The Glasgow Herald*, 17 December 1968.

9. https://dailytelegraph.co.nz ('The Volunteer Super-Spy: How a German Businessman Stole the Newest US Missile for Moscow', 27 November 2022).

Chapter Eight

1. https://nsarchive.gwu.edu ('NSC status report on Foreign Intelligence Program,' 28 July 1953).
2. Ibid. ('Memo from the Director of the Office of Research and Reports,' 13 May 1955).
3. Ibid. ('Sovmat Progress Report, 1 December 1955').
4. Ibid. ('Sovmat Staff Progress Report,' 1 December 1960).
5. Ibid. ('Memo from Lt Gen Joseph Carroll to Assistant Secretary of the Army,' 3 March 1966).
6. *Dogfights Episode 2: Air Ambush.*
7. *The Washington Post*, 6 August 1966.
8. Bar-Zohar and Mishal, *Mossad: The Great Operations of Israel's Secret Service*, p. 161.
9. Ibid. p. 164.
10. Ibid. p. 165.
11. Ibid. p. 166.
12. Ibid. p. 169.
13. Op. cit., https://nsarchive.gwu.edu ('Project Have Doughnut – Exploitation of the MiG-21', August 1969).
14. Ibid.
15. Op. cit., https://nsarchive.gwu.edu ('Project Have Drill/Have Ferry Tactical Evaluation Report', April 1970).
16. 'Aggressor' squadrons of the USAF simulated the tactics of the Warsaw Pact air forces in air combat exercises to improve the dogfight skills of US pilots.
17. The USAF bought two Sukhoi Su-27Ps from Belarus in 1995, while two more ex-Ukrainian Air Force examples were later purchased from a private US aviation company.

Chapter Nine

1. Geraghty, *Brixmis The Untold Exploits of Britain's Most Daring Cold War Spy Mission*, p. 330.
2. Wright and Jefferies, *Looking Down the Corridors*, p. 133.

3. Op. cit., *Brixmis*, p. 20.
4. Ibid. p. 84.
5. Ibid. pp. 240–41.
6. Op. cit., *Looking Down the Corridors*, p. 142.
7. Ibid. p. 132.
8. Op. cit., *Brixmis*, p. 84.
9. Op. cit., *Looking Down the Corridors*, p. 182.
10. Op. cit., *Brixmis*, p. 167.
11. *Proceedings* magazine, June 2018 issue ('Behind Enemy Lines: A Marine in East Germany' by Colonel Richard Camp USMC (Rtd.), published by the US Naval Institute).
12. National Archives, Kew: FO 1042/226.
13. https://nsarchive.gwu.edu ('President's Daily Brief,' 7 April 1966).
14. Op. cit., FO 1042/226.
15. Op. cit., nsarchive.gwu.edu ('President's Daily Brief,' 7 April 1966).
16. Op. cit., FO 1042/226.

Chapter Ten

1. Lashmar, *Spy Flights of the Cold War*, pp. 55–56.
2. Ibid. p. 56.
3. Ibid. p. 54.
4. *Smithsonian* magazine, July 2003.
5. Op. cit., *Spy Flights of the Cold War*, p. 55.
6. Mladenov, *Soviet Cold War Fighters*, p. 45.
7. https://en.wikipedia.org/wiki/K-13.
8. www.cia.gov/readingroom (CIA intelligence memorandum, 5 October 1965).
9. www.rbth.com ('How the Soviets fought against the Americans in Vietnam,' 7 July 2020).
10. Li, *The Dragon in the Jungle*, p. 145 (extracts reproduced with permission of the publisher © Oxford University Press).
11. Ibid. p. 146.
12. Price, *The History of US Electronic Warfare*, pp. 88–89.
13. *Journal of Military History*, Volume 67, January 2003 ('The -Ology War: Technology and Ideology in the Vietnamese Defense of Hanoi, 1967' by Merle L. Pribbenow).
14. Ibid.

15. https://nsrachive.gwu.edu ('Memo from Lt Gen Joseph Carroll to Deputy Secretary of Defense,' 5 Sept 1967, and 'Memo from Gen Robert Glass to Director of the Joint Staff,' 15 Nov 1967).
16. https://nsarchive.gwu.edu ('Memo from Lt Gen Eugene Tighe (USAF) to Deputy Under Secretary of Defense for Policy Affairs,' December 1980).
17. www.cia.gov/readingroom ('Afghanistan Situation Report', October 1985).
18. Geraghty, *Who Dares Wins*, p. 459.

Chapter Eleven

1. www.cia.gov/readingroom ('Soviet Supersonic: A Technopolitical Disaster,' December 1984).
2. Op. cit., cia.gov/readingroom ('Soviet Acquisition of Western Technology,' April 1982).
3. Andrew and Mitrokhin, *The Mitrokhin Archive*, p. 454.
4. Ibid. p. 282.
5. *New York Times*, 18 September 1976.
6. *Naval Aviation News*, February 1977.
7. *Newsweek*, 13 July 1981.
8. Op. cit., 'Soviet Acquisition of Western Technology'.
9. Op. cit., *The Mitrokhin Archive*, p. 543.
10. Ibid. p. 597.
11. *Bild*, 24 October 1984.
12. Weiss, *Duping the Soviets – the Farewell Dossier*.

Chapter Twelve

1. www.irp.fas.org.
2. www.cia.gov/readingroom ('Soviet Acquisition of Western Technology,' April 1982).
3. Ibid.
4. Ibid.
5. www.cia.gov/readingroom ('Soviet Acquisition of Militarily Significant Western Technology,' September 1985).
6. Ibid.

7. Kostin and Raynaud, *Farewell: The Greatest Spy Story of the Twentieth Century*, p. 172.

8. Op. cit., 'Soviet Acquisition of Militarily Significant Western Technology'.

9. Several theories have been advanced to explain how the KGB came to suspect Vetrov of treason, including the indiscreet letters he wrote in prison, a Soviet mole in a Western intelligence agency tipping off the KGB, or suspicions raised by the sudden wave of expulsions of Line X officers from foreign residencies in 1983.

10. Op. cit., *Farewell*, pp. 381-82.

11. Ibid. p. 333.

12. Ibid. p. 387.

13. Weiss, *Duping the Soviets – The Farewell Dossier*.

14. *US News & World Report*, 12 August 1985.

15. www.cia.gov/readingroom ('Soviet Supersonic: A Technopolitical Disaster,' December 1984).

16. *Christian Science Monitor*, 25 August 1980.

17. www.cia.gov/readingroom ('Soviet Work on Radar Cross Section Reduction Applicable to a Future Stealth Program,' February 1984).

18. *Washington Post*, 18 December 1984.

19. www.cia.gov/readingroom ('Soviet Reactions to Stealth,' 4 July 1985).

20. It wasn't until 2010 that a Russian combat aircraft with stealth qualities, the Sukhoi Su-57, first flew.

21. Mladenov, *Soviet Cold War Fighters*, p. 216.

22. Hoffman, *The Billion Dollar Spy*, p. 58.

23. www.cia.gov/readingroom ('Foxhound-A Dolinsk/Sokol airfield, USSR,' 7 November 1983).

24. Op. cit., *The Billion Dollar Spy*, pp. 147-48.

Chapter Thirteen

1. *Aviation News*, Vol. 85 No. 5, May 2023 issue ('Poking the Bear' by Peter R. Foster).

2. Richardson, *No Escape Zone*, p. 98.

3. *Los Angeles Times*, 8 July 1993.

4. How exactly the Serbs were able to track and shoot down Zelko's F-117 remains a matter of conjecture, with theories including that the opening

of the bomb bay doors or even the rain may have compromised the Nighthawk's radar invisibility being put forward as possible explanations.

5. www.assets.publishing.service.gov.uk (Defence Safety Authority Service Inquiry – 'Loss of F-35B Lightning ZM152 of 617 Squadron RAF').

6. www.nsa.gov.

7. TASS report, 6 July 2023.

8. https://en.wikipedia.org/wiki/Operation_Synytsia.

9. https://newsukraine.rbc.ua.

10. Op. cit., en.wikipedia.org/wiki/Operation_Synytsia. After his defection, Maxim Kuzminov relocated to the town of Villajoyosa on Spain's Costa Blanca but was found shot dead in an underground car park in February 2024. Inevitably, the SVR is suspected of being responsible for his murder.

11. www.globaldefensecorp.com ('Ukraine handed over R-37M air-to-air missile to British intelligence,' 21 February 2023).

12. www.fbi.gov.

13. www.bbc.co.uk/news.

14. *The EurAsian Times*, 16 March 2024.

15. www.osi.af.mil.

16. Reuters report, 19 June 2013.

17. *Defense News*, 24 March 2016.

18. *Express Tribune*, 16 August 2011.

19. *South China Morning Post*, 2 September 2019.

20. www.bbc.news.co.uk/news.

Conclusion

1. www.cia.gov/readingroom ('Soviet Supersonic: A Technopolitical Disaster').

2. Clarridge, *A Spy For All Seasons*, p. 154.

3. www.cia.gov/readingroom ('The Soviet Weapons Industry: An Overview,' September 1986).

4. Kostin and Raynaud, *Farewell: The Greatest Spy Story of the Twentieth Century*, p. 194.

5. Duffy and Kanalov, *Tupolev: The Man and his Aircraft*, p. 12.

6. The term applied to the battle waged by British and German scientists for technological superiority during the Second World War.

7. Bar-Zohar and Mishal, *Mossad: The Great Operations of Israel's Secret Service*, p. 168.

Bibliography

Books

Andrew, Christopher, *Defence of the Realm: The Authorized History of MI5* (Penguin, 2012)

Andrew, Christopher and Mitrokhin, Vasili, *The Mitrokhin Archive: The KGB in Europe and the West* (Penguin Books, 2000)

Ashcroft, Michael, *Heroes of the Skies* (Headline, 2013)

Bailes, Kendall, *Technology and Society Under Lenin and Stalin* (Princeton University Press, 1978)

Bar-Zohar, Michael and Mishal, Nissim, *Mossad: The Great Operations of Israel's Secret Service* (Biteback Publishing, 2015)

Brown, Captain Eric 'Winkle', *Wings on my Sleeve* (Weidenfeld & Nicholson, 2007)

Caygill, Peter, *Meteor from the Cockpit* (Pen & Sword Books, 2010)

Clarridge, Duane R., *A Spy For All Seasons* (Simon & Schuster, 2009)

Dancey, Peter G., *Soviet Aircraft Industry* (Fonthill Media, 2015)

Deane, John R., *The Strange Alliance* (Indiana University Press, 1971)

Drabkin, Artem, *Barbarossa and the Retreat to Moscow* (Pen & Sword, 2007)

Duffy, Paul and Kanalov, A.I., *Tupolev: The Man and his Aircraft* (SAE International, 1996)

Feklisov, Aleksandr and Kostine, Serguei, *The Man Behind the Rosenbergs* (Enigma, 2001)

Geraghty, Tony, *BRIXMIS The Untold Exploits of Britain's Most Daring Cold War Spy Mission* (HarperCollins, 1997)

__________, *Who Dares Wins: The Special Air Service – 1950 to the Gulf War* (Warner Books, 1993)

Gordon, Yefim and Rigmant, Vladimir, *OKB Tupolev A History of the Design Bureau and its Aircraft* (Midland Publishing, 2005)

______________, *Tupolev Tu-4 Soviet Superfortress* (Crecy Publishing, 2002)

Harden, Blaine, *King of Spies: The Dark Reign of America's Spymaster in Korea* (Pan MacMillan, 2017)

Hastings, Max, *The Secret War – Spies, Codes and Guerrillas 1939 – 45* (William Collins, 2015)

______________, *Vietnam: An Epic History of a Tragic War* (William Collins, 2019)

Haynes, John Earl, Klehr, Harvey and Vassiliev, Alexander, *Spies: The Rise and Fall of the KGB in America* (Yale University Press, 2009)

Hoffman, David E., *The Billion Dollar Spy* (Icon Books, 2018)

Holder, Bill, *Lost Fighters: A History of US Jet Fighter Programs That Didn't Make It* (SAE International, 2006)

Johnson, Loch K., *Secret Agencies: US Intelligence in a Hostile World* (Yale University Press, 1996)

Kostin, Sergei and Raynaud, Eric, *Farewell: The Greatest Spy Story of the Twentieth Century* (AmazonCrossing, 2011)

Lashmar, Paul, *Spy Flights of the Cold War* (Naval Institute Press, 1996)

Li, Xiaobing, *The Dragon in the Jungle: The Chinese Army in the Vietnam War* (Oxford University Press, 2020)

Lokhova, Svetlana, *The Spy Who Changed History: The Untold Story of how the Soviet Union Won the Race for America's Top Secrets* (William Collins, 2018)

Mladenov, Alexander, *Soviet Cold War Fighters* (Fonthill Media, 2016)

Moore, Jason Nicholas, *Soviet Strategic Bombers* (Fonthill Media, 2018)

Price, Alfred, *The History of US Electronic Warfare Volume III: Rolling Thunder Through Allied Force, 1964 – 2000* (Association of Old Crows, 2000)

Reshetnikov, Vasily, *Bomber Pilot of the Eastern Front* (Pen & Sword, 2008)

Richardson, Nick, *No Escape Zone* (Little Brown, 2000)

West, Nigel, *Soviet and Nazi Defectors – Counter-Intelligence in WW2 and the Cold War* (Pen & Sword Books, 2024)

______________, *The A to Z of British Intelligence* (Scarecrow Press, 2009)

______________, *MI5: British Security Service Operations 1909 – 1945* (Frontline, 2019)

Wright, Kevin and Jefferies, Peter, *Looking Down the Corridors: Allied Aerial Espionage over East Germany and Berlin 1945 – 1990* (The History Press, 2017)

Newspapers, magazines and journals

Aeroplane Monthly
Aviation News
Aviation Week
Bild
Christian Science Monitor
The EurAsian Times
The Express Tribune
FlyPast
Glasgow Herald
Journal of Military History
Los Angeles Times
Lodi News-Sentinel
Luftwaffe Secret Bombers of the Third Reich
MiG: Red Star Fighters
Naval Aviation News
New York Times
Newsweek
Philadelphia Enquirer
Proceedings magazine
Smithsonian magazine
South China Morning Post
US News & World Report
Washington Post

Archives and official documents

CIA reports:
Evolution of the Mikoyan design bureau since 1970s and MiG-25 (January 1990)
Foxhound-A, Dolinsk/Sokol airfield, USSR (November 1983)
Soviet Acquisition of Western Technology (April 1982)
Soviet Acquisition of Militarily Significant Western Technology (September 1985)
Soviet Reactions to Stealth (July 1985)
Soviet Supersonic: A Technopolitical Disaster (December 1984)
Soviet Work on Radar Cross Section Reduction Applicable to a Future Stealth Program (February 1984)

The Soviet Weapons Industry: an Overview (September 1986)

Venona: Soviet Espionage and the American Response 1939 – 1957 (October 1996)

Defence Safety Authority Service Inquiry – 'Loss of F-35B Lightning ZM152 of 67 Squadron RAF (17 November 2021)'

Defense Technical Information Center report: 'Duping the Soviets – the Farewell Dossier' by Gus Weiss (1996)

House Un-American Committee reports: Patterns of Communist Espionage (1959)

Soviet Espionage Activities in Connection with Jet Propulsion (June 1949)

National Archives (UK):

AIR 21/13016 ('Soviet aircraft landing on RAF airfields: policy')

AIR 22/93/1 ('Air Ministry Secret Intelligence Summary' January 1955)

AIR 27/2854/49 (No 237 Sqn Operations Record Book July 1954)

AIR 27/2388/3 (No 3 Sqn Summary of Events August 1948)

ADM 1/22578 ('Admiralty: Salvage of wrecked MiG-15 by HMS Glory off west coast of Korea, 1951')

CAB 79/15/24 ('No 30 Mission: Co-operation with the Russians', 3 November 1941)

CAB 79/23/35 (Chiefs of Staff Committee meeting, 10 October 1942)

CAB 79/82/10 ('Exchange of Technical Information with the USSR,' 27 October 1944)

CAB 128/26/18 ('Denmark: Disposal of MiG-15', 10 March 1953)

CAB 65/33/32 (Cabinet minutes, 18 February 1943)

DEFE 44/482 ('Translation of documents recovered from a Soviet Tu-16 Badger which

crashed into the Norwegian Sea' February 1969)

FO 371/94874 ('Enquiry about Russian Engineers employed at the Rolls-Royce factory in 1946')

FO 1042/226 ('Soviet air crash in Berlin' April – May 1966)

HW 15/25/22 ('New York: Information from "ARSENIJ" on jet aircraft engine' 12 August 1944)

HW 15/22/70 ('New York: GNOME's information on a long-distance fighter aircraft' 18 May 1944)

KV 2/1196 (Susan Benda)

KV 2/2029 and 2031 (Eric Camp)

KV 2/483 (Aleksey Doshenko)

KV 2/3059 (Benjamin Lockspeiser)

KV 2/647 (Wilfred Macartney)
KV 2/1024 (Thomas Mercer)
KV 2/220 and 2201 (Frederick Meredith)
KV 2/ 770 (James Messer)
KV 2/4600 (Olive Sheehan)
KV 2/1596 (Douglas Springhall)
KV 2/996 (Wilfred Vernon)
KV 2/643 (Leonid Vladimiroff)
KV 2/2199 (Ernest Weiss)
US Congressional Record, 16 May 1972

Documentaries

Dogfights Season One: Episode – 'Air Ambush' (The History Channel, 2006)
The MiG Story (Amity Productions, 1993)
Secret Superpower Aircraft – 'Spyplanes' (The History Channel, 2005)

Websites

www.archive.gov
www.bbc.co.uk/news
www.cia.gov/readingroom
https://dailytelegraph.co.nz
www.fbi.gov
www.globaldefensecorp.com
www.governmentattic.org
https://hansard.parliament.uk
https://history.state.gov
https://irp.fas.org
https://media.defense.gov
https://newsukraine.rbc.ua
https://nsarchive.gwu.edu
www.nsa.gov
www.osi.af.mil
https://people.uncw.edu
www.rbth.com
www.valor.militarytimes.com
https://en.wikipedia.org.uk

Index